AF593578

Fundamentals and Applications of Ternary Diffusion

Titles of Related Interest—

Ashby ENGINEERING MATERIALS 1
Ashby ENGINEERING MATERIALS 2
Brook IMPACT OF NON-DESTRUCTIVE TESTING
Koppel AUTOMATION IN MINING, MINERAL AND METAL PROCESSING 1989
Ruhle METAL-CERAMIC INTERFACES
Taya METAL MATRIX COMPOSITES

Other CIM Proceedings Published by Pergamon

Bergman FERROUS AND NON-FERROUS ALLOY PROCESSES
Bickert REDUCTION AND CASTING OF ALUMINUM
Bouchard PRODUCTION, REFINING, FABRICATION AND RECYCLING OF LIGHT METALS
Chalkley TAILING AND EFFLUENT MANAGEMENT
Closset PRODUCTION AND ELECTROLYSIS OF LIGHT METALS
Dobby PROCESSING OF COMPLEX ORES
Embury HIGH TEMPERATURE OXIDATION AND SULPHIDATION PROCESSES
Jaeck PRIMARY AND SECONDARY LEAD PROCESSING
Jonas DIRECT ROLLING AND HOT CHARGING OF STRAND CAST BILLETS
Kachaniwsky IMPACT OF OXYGEN ON THE PRODUCTIVITY OF NON-FERROUS METALLURGICAL PROCESSES
Lait F. WEINBERG INTERNATIONAL SYMPOSIUM ON SOLIDIFICATION PROCESSING
Macmillan QUALITY AND PROCESS CONTROL IN REDUCTION AND CASTING OF ALUMINUM AND OTHER LIGHT METALS
Mostaghaci PROCESSING OF CERAMIC AND METAL MATRIX COMPOSITES
Plumpton PRODUCTION AND PROCESSING OF FINE PARTICLES
Rigaud ADVANCES IN REFRACTORIES FOR THE METALLURGICAL INDUSTRIES
Ruddle ACCELERATED COOLING OF ROLLED STEEL
Salter GOLD METALLURGY
Thompson COMPUTER SOFTWARE IN CHEMICAL AND EXTRACTIVE METALLURGY
Twigge-Molecey MATERIALS HANDLING IN PYROMETALLURGY
Twigge-Molecey PROCESS GAS HANDLING AND CLEANING
Tyson FRACTURE MECHANICS
Wilkinson ADVANCED STRUCTURAL MATERIALS

Related Journals

(Free sample copies available upon request)

ACTA METALLURGICA
CANADIAN METALLURGICAL QUARTERLY
MATERIALS RESEARCH BULLETIN
MINERALS ENGINEERING
SCRIPTA METALLURGICA

PROCEEDINGS OF THE INTERNATIONAL SYMPOSIUM ON FUNDAMENTALS AND APPLICATIONS OF TERNARY DIFFUSION HAMILTON, ONTARIO, CANADA, AUGUST 27-28, 1990

Fundamentals and Applications of Ternary Diffusion

Editor
G.R. Purdy
Department of Materials Science and Engineering
McMaster University
Hamilton, Ontario, Canada

Symposium organized by The Metallurgical Society of CIM

29th ANNUAL CONFERENCE OF METALLURGISTS OF CIM
29e CONFÉRENCE ANNUELLE DES MÉTALLURGISTES DE L'ICM

Pergamon Press
Member of Maxwell Macmillan Pergamon Publishing Corporation
New York Oxford Beijing Frankfurt São Paulo Sydney Tokyo Toronto

Pergamon Press Offices:

U.S.A.	Pergamon Press, Inc., Maxwell House, Fairview Park, Elmsford, New York 10523, U.S.A.
U.K.	Pergamon Press plc, Headington Hill Hall, Oxford OX3 0BW, England
PEOPLE'S REPUBLIC OF CHINA	Pergamon Press, 0909 China World Tower, No. 1 Jian Guo Men Wai Avenue, Beijing 1000004, People's Republic of China
FEDERAL REPUBLIC OF GERMANY	Pergamon Press GmbH, Hammerweg 6, D-6242 Kronberg, Federal Republic of Germany
BRAZIL	Pergamon Editora Ltda, Rua Eça de Queiros, 346 CEP 04011, Paraiso, São Paulo, Brazil
AUSTRALIA	Pergamon Press Australia Pty Ltd., P.O. Box 544, Potts Point, NSW 2011, Australia
JAPAN	Pergamon Press, 8th Floor, Matsuoka Central Building, 1-7-1 Nishishinjuku, Shinjuku-ku, Tokyo 160, Japan
CANADA	Pergamon Press Canada Ltd., Suite 271, 253 College Street, Toronto, Ontario M5T 1R5 Canada

Library of Congress Cataloging in Publication Data

ISBN 0-08-040412-X

Printing: 1 2 3 4 5 6 7 8 9 Year: 0 1 2 3 4 5 6 7 8 9

Printed in the United States of America

The paper used in this publication meets the minimum requirements of American National Standard for Information Sciences-Permanence of Paper for Printed Library Materials, ANSI Z 39.48-1984

Foreword

In 1955, as a junior professor at McGill University and a summer employee of Aluminium Laboratories in Kingston, Ontario, Jack Kirkaldy encountered the problems of diffusion and phase transformations in multicomponent alloys and as a physicist-cum-metallurgist decided to undertake a systematic experimental and theoretical study of the phenomena with a view to establishing a technological problem-solving capability. Upon joining McMaster University in 1957, he established a large and productive research group, focussing on aluminum and iron-carbon base solid alloys, but not excluding the full range of solid and liquid material manifestations. His synthesis, with D.J. Young, appeared in 1987 as the definitive volume on multicomponent effects, Diffusion in the Condensed State. The present volume may be regarded as an appendix and update on that synthetic contribution.

Taking a lead from Onsager's irreversible thermodynamic foundations, Kirkaldy, in 1958, recognized the importance of matrix methods in generalizing binary theory to ternary and higher order systems. Elements of this intelligence are to be found in the new contributions of Miller, Morral *et al.* and Kirkaldy *et al.* in this volume. Jack recognized early on that ternary cross effects would often be weak, as for example in dilute solutions, so that pseudo-binary constructions would be adequate for problem-solving. While establishing appropriate theoretical criteria, he also sought to discover material or experimental conditions where such cross-effects would be enhanced. Papers in the present volume co-authored with Young, Okongwu, Zou and Buchmayr expand this particular knowledge to a range of new material systems.

There is a broad trend toward interaction enhancement at concentrations rich enough to produce phase transformations. Kirkaldy and co-workers have accordingly extensively studied the multicomponent-multiphase phenomena in both stable and unstable interface configurations. This work in Fe-C-X alloys has had an important technological impact through the successful marketing of Jack's software package pertaining to quality control and optimization in the heat treatment of alloy steels. The papers of Dayananda and Liu, Jan *et al.*, van Loo *et al.*, Stringer, Brechet, Balasubramanian, Zou, Larson and Buchmayr bear directly on this problem area and exploit Jack's earlier contributions.

In 1963-66, Lane and Kirkaldy demonstrated that the Kirkendall Effect associated with a vacancy diffusion mechanism in solids produces a ternary diffusion cross-effect through the drift velocity which can be comparable in magnitude to purely thermodynamic effects. In his present short contribution, he makes an explicit comparison of magnitudes and demonstrates that vacancy correlation effects on diffusion profiles are usually unimportant in close-packed materials.

In early studies of unstable ternary transformation interfaces, Jack recognized a close analogy with instabilities generated in cellular solidification of binary alloys in a temperature gradient. This has led him into a deep study of pattern formation based within the variational principles of irreversible thermodynamics and formal logic. While excluded from this volume, the elements are to be found in Chapter 13 of the above-mentioned 1987 volume and the Pergamon symposium volume "Advances in Phase Transformations", which was arranged to celebrate Jack's 60th birthday. With this volume, presented upon his retirement, we gratefully acknowledge Jack Kirkaldy's many contributions to, and intellectual leadership in, the field of ternary diffusion over more than three decades.

Gary Purdy
Hamilton, Ontario

August 1990

Table of Contents

3. APPLICATIONS

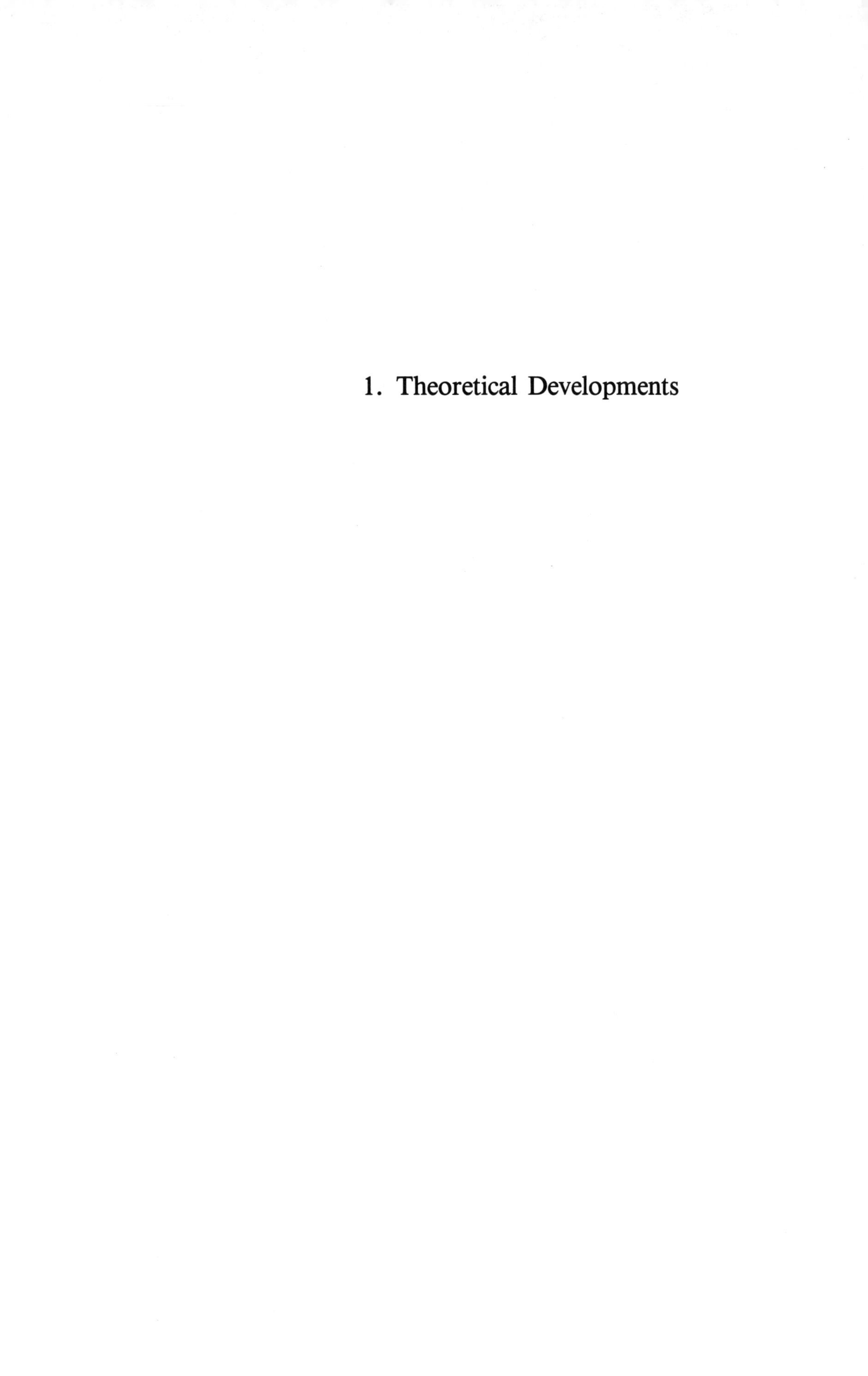

1. Theoretical Developments

Impurity diffusion in a doubly-doped elemental semiconductor with pair formation

D.J. Young
School of Materials Science and Engineering, University of New South Wales, Kensington, N.S.W. 2033, Australia

R.C. Dorward
Kaiser Chemical and Aluminum Corp., Pleasanton, California, U.S.A.

J.S. Kirkaldy
Institute for Materials Research, McMaster University, Hamilton, Ontario, Canada, L8S 4M1

Abstract

The diffusion of substitutionally dissolved donors and acceptors has been described in terms of point defect movement and an irreversible thermodynamic treatment has been used to identify the interactions among the different diffusing species. The principal diffusive interactions for a donor species have been shown to be countercurrent donor-vacancy movement, countercurrent donor-solvent movement, countercurrent donor-acceptor movement and co-current donor-acceptor movement. The relative magnitude of kinetic coefficients can be calculated specifically from an absolute rate theory approach. This paper deals specifically with the common case of countercurrent donor-acceptor movement subject to the complication of donor-acceptor pair formation. This is incorporated into the kinetics using a linear expansion of the logarithm of the activity coefficient. A representative diffusion profile is given showing significant attractive segregation of the donor on a gradient of acceptor despite the extreme diluteness of the reactants.

Introduction

Diffusion against a concentration gradient in a ternary or higher order couple can occur for a variety of reasons including thermodynamic interactions, Onsager correlation cross-effects, vacancy winds and inhomogeneities, electrostatic neutralization interactions in ionic solids (1) and electron-solute interactions in semiconductors (2). This paper deals with the exceptionally strong effect of an electrostatic pair reaction (1-3) as represented by combination of donor and acceptor dopants in a semiconductor (e.g. Si). At the very dilute concentrations involved thermodynamic solution and Onsager effects (1) are by contrast of no observational consequence.

An elemental semiconductor such as silicon will in general contain a variety of defect types, including impurities. In describing the diffusion of those impurities, it is necessary to take into account their chemical interactions with all other species in the solid. An appropriate methodology for calculating these effects is provided by irreversible thermodynamics, which provides a phenomenological relationship between fluxes and driving forces and interrelates the different species. For any system of Schottky and (or) Frenkel defects we adopt the notation of Kroger and Vink (4) wherein the lattice species are represented by the symbol S^x_M. Here the subscript represents the normal occupancy in a perfect crystal of the site in question, and the principal symbol represents the species actually occupying the site. The superscript represents the charge of the species relative to normal site occupancy with a prime denoting a negative, a dot positive and a cross, zero charge.

When defects diffuse under isothermal, field-free conditions, they must on the average move in groups which conserve both electrical charge and sites. These groups satisfy the definition of "relative building units" formulated by Schottky (5) and by Kroger et al. (6) in their description of point defect equilibria. The usefulness of this concept in treating diffusion in ionic solids has since been demonstrated (7, 8). Kirkaldy and Young (1) have adapted that treatment to the case of an elemental semiconductor containing multiple vacancy charge states, as well as impurities. They have also discussed procedures for introducing the diffusion kinetics via absolute rate theory, and this will be expanded in the present context of diffusing building units. Finally, the effect of donor-acceptor pair formation on the observed diffusion behaviour is analyzed. Although the treatment is quite general, the specific case of silicon is used as an example.

The Relative Building Units For A Doubly-Doped Semiconductor

This section summarizes the results of the irreversible thermodynamic entropy source method for identifying the building units U_i (1). In this method, the linear dependencies contained in the set of fluxes appearing in the entropy source term are eliminated using the flux constraints due to site and charge conservation.

For silicon containing a donor, D, an acceptor, A, and vacancies which can be positively or negatively charged, as well as neutral, there exist 18 distinguishable pairs of species fluxes not involving the donor. Elimination of these flux pairs reveals the building units containing D, in serial order (i = 1-13):

$$
\begin{array}{l}
\{D^{\cdot}_{Si} - V^{\prime}_{Si}\} \quad \{D^{\cdot}_{Si} - V^{x}_{Si} - h^{\cdot}\} \quad \{D^{\cdot}_{Si} - V^{x}_{Si} + e^{\prime}\} \quad \{D^{\cdot}_{Si} - 2V^{x}_{Si} + V^{\prime}_{Si}\} \\
D_{Si} - V^{\prime}_{Si} + 2e^{\prime}\} \quad \{D^{\cdot}_{Si} - V^{\prime}_{Si} + 2h^{\cdot}\} \\
\{D^{\cdot}_{Si} - Si^{x}_{Si} + e^{\prime}\} \quad \{D^{\cdot}_{Si} - Si^{x}_{Si} - h^{\cdot}\} \quad \{D^{\cdot}_{Si} - 2Si^{x}_{Si} + V^{\prime}_{Si}\} \\
\{D^{\cdot}_{Si} - A^{\prime}_{Si} + 2e^{\prime}\} \quad \{D^{\cdot}_{Si} - A^{\prime}_{Si} - 2h^{\cdot}\} \\
\{D^{\cdot}_{Si} + A^{\prime}_{Si} - 2V^{x}_{Si}\} \quad \{D^{\cdot}_{Si} + A^{\prime}_{Si} - 2Si^{x}_{Si}\}
\end{array}
\tag{1}
$$

Only 13 of 18 distinguishable units result because all six units involving $V^{\cdot}_{Si}$ reduce to $\{D^{\cdot}_{Si} - V^{\cdot}_{Si}\}$. It should be noted that these units reflect the simplest forms of correlation, as any linear combination of the units would obviously satisfy all thermodynamic and conservative requirements.

The relative building units involving $D^{\cdot}_{Si}$ necessarily reflect the flux constraints which via the entropy source expression impose the requirements that movement of $D^{\cdot}_{Si}$ be balanced by an opposite movement of another lattice species, and that any net charge flow thereby arising be neutralised. Thus four principal correlations are seen to exist; countercurrent donor-vacancy movement, countercurrent donor-solvent movement, countercurrent donor-acceptor movement and cocurrent donor acceptor movement.

The relative importance of these various possible correlations will be determined by the probabilities of finding the required species in adjacent positions within the solid (i.e., of locating the required relative building unit). Thus the most likely of the various possible donor-vacancy exchange mechanisms will be determined by the vacancy ionization equilibria. All correlations involving the simultaneous movement of two constituents require the presence of a vacancy. If the vacancy is explicit in the corresponding relative building unit, then net vacancy transport occurs. If the unit contains no vacancy, then the vacancies function only as intermediaries. It follows then that donor-solvent countercurrent diffusion can occur with and without net vacancy movement. The favoured mechanism will depend on the relative abundance of V'_{Si} and V^x_{Si}. Cocurrent donor-acceptor diffusion can also occur both with and without a vacancy flux. Given that the probability of finding a Si^x_{Si} species approaches unity, the unit $\{D^{\cdot}_{Si} + A'_{Si} - 2Si^x_{Si}\}$ adjacent to an additional V^x_{Si} is more probable than $\{D^{\cdot}_{Si} + A'_{Si} - 2V^x_{Si}\}$ alone. The zero vacancy flux situation therefore seems more probable. Countercurrent donor-acceptor diffusion cannot involve a net vacancy flux but must involve an electronic current. The relative importance of cocurrent and countercurrent donor-vacancy correlations will depend on kinetic factors and on the relative availability of neutral vacancies and free carriers.

Applying the same procedures to an analysis of the diffusion of acceptors A′ it is found that the relative building units U'_i involved are in line serial order (i = 1-13)

$$
\begin{gathered}
\{A'_{Si} - V'_{Si}\}\ \{A'_{Si} - V^x_{Si} + h^{\cdot}\}\ \{A'_{Si} - V^x_{Si} - e'\}\ \{A'_{Si} - 2V^x_{Si} + V^{\cdot}_{Si}\} \\
\{A'_{Si} - V^{\cdot}_{Si} - 2e'\}\ \{A'_{Si} - V^{\cdot}_{Si} + 2h^{\cdot}\} \\
\{A'_{Si} - Si^x_{Si} - e'\}\ \{A' - Si^x_{Si} + h^{\cdot}\}\ \{A'_{Si} - 2Si^x_{Si} + V^{\cdot}_{Si}\} \\
\{A'_{Si} - D^{\cdot}_{Si} - 2e'\}\ \{A'_{Si} - D^{\cdot}_{Si} + 2h^{\cdot}\} \\
\{A'_{Si} + D^{\cdot}_{Si} - 2V^x_{Si}\}\ \{A'_{Si} + D^{\cdot}_{Si} - 2Si^x_{Si}\}
\end{gathered}
\tag{2}
$$

The mechanism by which $D^{\cdot}$ can move is via the relative building units listed in Eq. 1. A linear combination of the fluxes arising therefrom represents a complete and exhaustive accounting for diffusion of $D^{\cdot}$, viz.,

$$J_D = -\sum_{i=1}^{13} l_i \nabla\eta\{U_i\} \tag{3}$$

where $\eta(U_i)$ is the electrochemical potential of building unit U_i of standard form

$$\eta_i = \mu_i + z_i F \psi \tag{4}$$

where μ_i is the chemical potential, z_i is the number of electronic charges with due regard to sign, F is the Faraday and ψ is the electrochemical potential. As usual ψ vanishes in the transformations. Substituting for $\eta(U_i)$ from Eq. 1 one obtains after rearrangement

$$J_D = -\sum_{i=1}^{13} l_i \nabla\mu_D - (l_{12} + l_{13} - l_{10} - l_{11}) \nabla\mu_A + (l_7 + l_8 + 2l_9 + 2l_{13}) \nabla\mu_{Si} \tag{5}$$

and the cross-terms in the diffusion matrix are seen to be explicitly identified.

Similarly, it is found for diffusion of A′

$$J_A = -(\lambda_{12} + \lambda_{13} - \lambda_{10} - \lambda_{11}) \nabla\mu_D - \sum_{i=1}^{13} \lambda_i \nabla\mu_A + (\lambda_7 + \lambda_8 + 2\lambda_9 + 2\lambda_{13}) \nabla\mu_{Si} \tag{6}$$

Further simplification results if the solid solution is Henrian, for then in terms of concentration m_i in moles/unit volume

$$\nabla\mu_i = RT\ \nabla m_i / m_i \tag{7}$$

and, for low vacancy concentrations,

$$\nabla m_{Si} = -\nabla m_D - \nabla m_A \tag{8}$$

Substitution of (7) and (8) into (5) and (6) followed by rearrangement yields

$$\frac{J_D}{RT} = -\left\{\frac{L_{DD}}{m_D} - \frac{L_{DSi}}{m_{Si}}\right\}\nabla m_D - \left\{\frac{L_{DA}}{m_A} - \frac{L_{DSi}}{m_{Si}}\right\}\nabla m_A$$

$$\frac{J_A}{RT} = -\left\{\frac{L_{AD}}{m_D} - \frac{L_{ASi}}{m_{Si}}\right\}\nabla m_D - \left\{\frac{L_{AA}}{m_A} - \frac{L_{ASi}}{m_{Si}}\right\}\nabla m_A \tag{9}$$

where the L_{ij} are defined as the coefficients in Eqns. (5) and (6).

To proceed further, the as-yet unidentified coefficients ℓ_i and λ_i must be evaluated. To this end we must turn to a kinetic description of the process. We may anticipate, however, that because the units U_{10} to U_{13} of Eq. 1 involve precisely the same lattice and free carrier species as the analogous units U'_{10} to U'_{13} of Eq. 2, the kinetics of the species interchange processes will be the same, and hence

$$l_i = \lambda_i \ , \quad i = 10, 11, 12, 13 \tag{10}$$

In consequence of this eventuality, the symmetry expected of the L-matrix is arrived at.

Defect Diffusion Kinetics

The diffusion of point defects can be calculated by the methods of absolute rate theory (9) when proper account is taken of local electrostatic field effects (10). The method has been presented elsewhere (11). For the interchange of lattices species i and vacancies between adjacent sites separated by a distance d

$$J = -\frac{Nd^2}{RT}\nu_{iv} K_{iv} a_i a_v \nabla\{\eta_i - \eta_v\} \tag{11}$$

Here N is the volume concentration of lattice sites, ν_{iv} is a frequency term, and K_{iv} is the equilibrium constant relating the activity of the transition-state complex to those of the reactants (a_i and a_v). This expression is used to evaluate the rates of the several pair-wise interchanges which constitute the movement of a complete relative building unit.

Applying Eqn. (11) to the diffusion of U_{10}, for example, we find

$$J^{(a)} = -\frac{Nd^2}{RT}\nu_{DV^x} K_{DV^x} a_D a_{V^x} \nabla\{\eta(D^{\cdot}) - \eta(V^x)\} \tag{12}$$

$$J^{(b)} = -\frac{Nd^2}{RT}\nu_{AV^x} K_{AV^x} a_A a_{V^x} \nabla\{\eta(V^x) - \eta(A^{'})\} \tag{13}$$

$$J^{(c)} = J^{(d)} = -\frac{Nd^2}{RT}\nu_e K_e a_e \nabla\{\eta(e^{'})\} \tag{14}$$

In addition, it is possible to write for the overall process from initial to final state

$$J_{10} = -\frac{Nd^2}{RT}\nu K_C a_D a_A a_{V^x} a_e^2 \nabla\{\eta(D^{\cdot}) - \eta(A^{'}) + 2\eta(e^{'})\} \tag{15}$$

The quantity νK_C is evaluated from the quasi-steady-state condition

$$J = J^{(a)} = J^{(b)} = J^{(c)}$$

and the combined activity product corresponding to the four reactions of Eqs. 12-14

$$K_C = K_{D'V}x\, K_{A'V}x\, K_e^2 \quad (16)$$

It is then found that

$$J_{10} = -\frac{Nd^2}{RT}\,\frac{m_D' m_A' m_e'}{m_A' m_e' + m_D' m_e' + 2m_D' m_A' a_V x}\, a_V x \nabla\{\eta(D') - \eta(A') + 2\eta(e')\} \quad (17)$$

where new concentration variables are defined by

$$m_D' = a_D \nu_{DV} x\, K_{DV}\ ;\quad m_A' = a_A \nu_{AV} x\, K_{AV} x\ ;\quad m_e' = a_e \nu_e K_e \quad (18)$$

The coefficient l_{10} is thus evaluated. Values found in this way for all the l_i, λ_i are listed in the Appendix. It is seen here explicitly that

$$l_i \neq \lambda_i\ ,\quad i = 1, 2 \ldots 9$$
$$l_i = \lambda_i\ ,\quad i = 10, \ldots 13 \quad (19)$$

and the L-matrix is symmetric.

The relative importance of the different diffusion mechanisms depends upon the concentrations of the individual species involved which in turn depend on the equilibrium constants for the formation reactions (1, 12). Even though all relative building units, when they move, represent diffusion without a net current, several of them contain a net charge, e.g., $\{D^{\cdot}_{Si} - V^{\cdot}_{Si}\}$. Movement of these units obviously involves two equal and opposite electrical currents. However, the existence of sufficient units to provide a diffusive flux necessarily implies a space charge unless local compensating, oppositely charged defects are present. This, in turn, implies that the probability of finding the required species in adjacent positions within the solid (i.e. of locating the required relative building unit) is low. In other words, because they possess mobile free carriers, semiconductors cannot support large space charges and therefore the concentration of uncompensated building units is necessarily low. Accordingly, diffusion contributions from the following units are taken to be negligible:

$$U_1,\ U_2,\ U_5,\ U_6,\ U_8,\ U_{10},\ U_{11},$$

$$U_1',\ U_3',\ U_5',\ U_6',\ U_7',\ U_{10}',\ U_{11}'$$

The contributions of the remaining units are now used to evaluate the coefficients in the diffusion equations (9). It is convenient to recast these as

$$\frac{J_D}{RT} = -\frac{\nabla m_D}{m_D}\left\{L_{DD} - L_{DSi}\frac{m_D}{m_{Si}}\right\} - \frac{\nabla m_A}{m_A}\left\{L_{DA} - L_{DSi}\frac{m_A}{m_{Si}}\right\}$$
$$\frac{J_A}{RT} = -\frac{\nabla m_D}{m_D}\left\{L_{AD} - L_{ASi}\frac{m_A}{m_{Si}}\right\} - \frac{\nabla m_A}{m_A}\left\{L_{AA} - L_{ASi}\frac{m_A}{m_{Si}}\right\} \quad (20)$$

Noting from definitions (18) that for Henrian behaviour

$$\nabla m_i/m_i = \nabla m_i'/m_i' \quad (21)$$

and substituting for the L_{ij} from term-by-term comparison with Eqns. (5) and (6) then, if only charge-neutral relative building units are considered, we obtain (dropping the primes for clarity)

$$J_D = -D_{DD}\nabla m_D - D_{DA}\nabla m_A$$
$$J_A = -D_{AD}\nabla m_D - D_{AA}\nabla m_A \qquad (22)$$

$$\frac{D_{DD}}{Nd^2} = a_V x\left\{\frac{2m_e}{m_e + m_D a_V x} + \frac{2m_A}{m_A + m_D} + \frac{m_{V'}}{m_{V'} + m_D}\right\} + a_{V'}$$

$$\frac{D_{DA}}{Nd^2} = a_V x\left\{\frac{2m_D}{m_A + m_D} + \frac{m_D m_D m_e}{m_{Si}\, m_A(m_e + m_D a_V x)}\right\} + 2\,\frac{m_D}{m_{Si}}\,\frac{m_D}{m_A}\,a_{V'}$$

$$\frac{D_{AD}}{Nd^2} = a_V x\left\{\frac{2m_A}{m_A + m_D} + \frac{m_D m_A m_h}{m_{Si}\, m_D(m_h + m_A a_V x)}\right\} + 2\,\frac{m_D}{m_{Si}}\,\frac{m_A}{m_D}\,a_{V\cdot} \qquad (23)$$

$$\frac{D_{AA}}{Nd^2} = a_V x\left\{\frac{2m_h}{m_h + m_A a_V x} + \frac{2\,m_D}{m_A + m_D} + \frac{m_{V'}}{m_{V'} + m_A}\right\} + a_{V\cdot}$$

Note that the m's in Eqs. 22 as defined incorporate factors of form νK.

Consequences of Defect Pairing

Point defects may be able to interact strongly with one another (1-3, 13) forming complexes which are relatively immobile within the host lattice. An example of this additional kind of local equilibrium is pair formation between two oppositely charged donors and acceptors in a semiconductor represented by the chemical equation

$$D^{\cdot} + A' = DA \qquad (24)$$

for which we may write the mass law, assuming ideal or Henrian solution behaviour

$$X_{DA}/X_{D^{\cdot}}\,X_{A'} = K \qquad (25)$$

In this case the concentrations of species actually free to move, $X_{D^{\cdot}}$ and $X_{A'}$, differ from the analytically measured concentrations of these components, X_1 and X_2, i.e.

$$X_D = X_1 - X_{DA} \qquad X_{A'} = X_2 - X_{DA} \qquad (26)$$

Since diffusion experiments on solids invariably rely on analytical measurements of total component concentration, the diffusion equations must be couched in these terms. It is necessary, therefore, to describe the thermodynamics of the solution in terms of total component concentrations.

Whilst the free species $D^{\cdot}$ and A', if dilute, will be Henrian, it is most improbable that the solute components X_1 and X_2 can be so regarded. This is an immediate consequence of the fact that X_1, X_2 are not direct measures of $X_{D^{\cdot}}$ and $X_{A'}$ because of the formation of pairs. Consequently activities referred to the pure solutes are expected to deviate markedly from their values in the absence of pair formation. One particular solution model is now proposed and its consequences for the diffusion equations explored.

If the two solutes have negligible solubility for the solvent and each other, then the phase boundaries of the solution may be defined, at constant temperature and constant activity lines (Fig. 1).

$$\gamma_1 X_1 = 1 \quad \text{and} \quad \gamma_2 X_2 = 1 \qquad (27)$$

for, respectively, the solution-component 1 and solution-component 2 boundaries. If the further condition is applied that unassociated solutes display Henrian solution behaviour, then, for Henrian activity coefficients $h_{A'}$, $h_{D^{\cdot}}$,

$$h_{D^{\cdot}} X_{D^{\cdot}} = 1 \quad \text{and} \quad h_{A'} X_{A'} = 1 \tag{28}$$

from which it follows that the phase boundaries are defined by constant concentrations of these species

$$X_{D^{\cdot}} = X^0_{D^{\cdot}} \quad \text{and} \quad X_{A'} = X^0_{A'} \tag{29}$$

respectively. The component activity coefficients γ_1, γ_2 can now be calculated.

Substitution from equations (26) and (29) into equation (25) followed by rearrangement yields for the pair concentration

$$X_{DA} = \frac{KX^0_D}{1 + KX^0_D} X_2 \tag{30}$$

which applies along the solution-component 1 boundary. The concentration of X_1 along this boundary is then found from equations (26) and (30) as

$$X_1 = X^0_D + \frac{KX^0_D}{1 + KX^0_D} X_2 \tag{31}$$

whence, through Eq. (27)

$$\ln \gamma_1 \simeq \left(\frac{1 + KX^0_D}{X^0_D} \right) - KX^0_D - KX_2 \tag{32}$$

Similarly, it is easily shown that along the solution-component 2 boundary

$$\ln \gamma_2 \simeq \ln \left(\frac{1 + KX^0_A}{X^0_A} \right) - KX^0_D - KX_1 \tag{33}$$

These phase boundary results are seen to continue analytically to Wagner Taylor expansion expressions (14) for activity coefficients in dilute ternary solutions, namely

$$\begin{aligned} \ln \gamma_1 &= K_1 + \beta_1 X_1 + \beta X_2 \\ \ln \gamma_2 &= K_2 + \beta X_1 + \beta_2 X_2 \end{aligned} \tag{34}$$

with $\beta_1 = \beta_2 = 0$; $\beta = -K$. The vanishing of the on-diagonal coefficients is a consequence of the foregoing Henrian assumption which neglects self-interactions relative to the ionic cross-interactions.

The equilibrium constant is related to the binding free energy of the complex by (2, 3, 13)

$$K = \exp(\Delta F/kT) \tag{35}$$

If we neglect the entropy of formation we can write this as

$$K \simeq \exp(\Delta E/kT) \tag{36}$$

where ΔE can be approximated by the binding energy of the complex. This can be typically estimated as the lattice energy of a monovalent ionic solid (~7ev) divided by the dielectric constant (12 for Si and 16 for Ge). Thus for a typical diffusion temperature range, we may expect

$$10^4 < K < 10^7 \tag{37}$$

Thus, via Eqs. 34 and 37, it can be appreciated how even small doping concentrations can have a large effect on solubility and on diffusion behaviour.

Diffusion in the Presence of Pairing

For the pairing or complexing situation, we must refer to components 1 and 2 (which are identical to $D^{\cdot}$ and A', respectively, in the absence of pairing). Equations 5 and 6 still hold with $\mu_D \rightarrow \mu_1$, $\mu_A \rightarrow \mu_2$. Then we evaluate $\nabla\mu_i$ from the pair thermodynamics model (Eqs. 34) giving in terms of mole fractions X_i

$$\nabla\mu_1 = RT\frac{\nabla X_1}{X_1} + RT\,\nabla \ln \gamma_1 = RT\left\{\frac{\nabla X_1}{X_1} - K\,\nabla X_2\right\}$$

$$\nabla\mu_2 = RT\left\{\frac{\nabla X_2}{X_2} - K\,\nabla X_1\right\} \tag{38}$$

Substitution of these into Eqs. 5 and 6 yields

$$\frac{J_1}{RT} = -\left\{\frac{m_D D_{DD}}{X_1} - m_A D_{DA} K\right\}\nabla X_1 - \left\{\frac{m_A D_{DA}}{X_2} - m_D D_{DD} K\right\}\nabla X_2$$

$$\frac{J_2}{RT} = -\left\{\frac{m_D D_{AD}}{X_1} - m_A D_{AA} K\right\}\nabla X_1 - \left\{\frac{m_A D_{AA}}{X_2} - m_D D_{AD} K\right\}\nabla X_2 \tag{39}$$

which we summarize as

$$J_1 = -D_{11}\nabla(M X_1) - D_{12}\nabla(M X_2) = -D_{11}\nabla m_1 - D_{12}\nabla m_2$$

$$J_2 = -D_{21}\nabla(M X_1) - D_{22}\nabla(M X_2) = -D_{21}\nabla m_1 - D_{22}\nabla m_2 \tag{40}$$

where $M = m_1 + m_2 + m_{Si}$. This is the appropriate modification of Eqs. 22.

We next undertake the evaluation of the D-matrix of Eqs. 40 explicitly for a typical limiting case, e.g. when

$$m_{Si} >> m_D >> m_A \tag{41}$$

for the common case of extrinsic behaviour, viz.,

$$m_{D^{\cdot}} = m_{e'}; \qquad m_{h^{\cdot}}\, m_{e'} = \text{constant} \tag{42}$$

and the particular case where neutral vacancies are dominant, i.e.

$$m_{V^{\cdot}} \simeq 0 \simeq m_{V'} \tag{43}$$

and where as a consequence of (41) the hole concentration is relatively small, i.e.,

$$m_{h^{\cdot}} << m_{e'} \tag{44}$$

These conditions are determined by lattice-defect equilibria which have not been introduced explicitly here (1,12). We also assume symmetry within the transition state matrices of the form (cf. Appendix)

$$Nd^2 RT\,\nu_{DV}\,K_{DV} \simeq Nd^2 RT\,\nu_{AV}\,K_{AV} = \kappa \tag{45}$$

Now recognizing in Eqs. 25 et seq. that $X_i = m_i/M$ we substitute for m_A, m_D in terms of m_1, m_2

$$m_D = m_1 - m_{DA} = \frac{1}{2}\left(m_1 - m_2 - \frac{M}{K}\right) + \left[\frac{1}{4}\left(m_2 - m_1 + \frac{M}{K}\right)^2 + m_1\frac{M}{K}\right]^{1/2}$$

$$m_A = m_2 - m_{DA} = \frac{1}{2}\left(m_2 - m_1 - \frac{M}{K}\right) + \left[\frac{1}{4}\left(m_1 - m_2 + \frac{M}{K}\right)^2 + m_2\frac{M}{K}\right]^{1/2} \quad (46)$$

These considerations lead finally to the Fick equations

$$\kappa' J_1 = -2(M - m_A K)\nabla m_1 - 2(M - m_D K)\nabla m_2$$

$$\kappa' J_2 = 2\, m_A K \nabla m_1 - 2M\nabla m_2 \quad (47)$$

where

$$\kappa' = \frac{M}{a_V x\, \kappa} \quad (48)$$

where m_D and m_A are given by Eqs. 46. For the diffusion matrix we have via Eq. 25 the trace

$$\mathrm{tr}\, D_{ik} = M(4 - 2m_{DA}/m_A)/\kappa' \quad (49)$$

and the determinant

$$\det D_{ik} = 4(M^2 - m_{DA} K M)/\kappa'^2 \quad (50)$$

Since $m_{DA} < m_A$ the trace is always positive. However, the determinant can change sign if κ is large enough relative to the leanest of the two solute additions indicating a solution instability (see below).

From the various conditions previously applied, and specifically letting magnitudes $m_1 = 2\times10^{-4}$ M, $m_2 = 5\times10^{-6}$ M, which are typically dilute and $K = 10^5$ we can approximate to obtain the defect concentrations

$$m_D = m_1 \quad (51)$$

and

$$m_A = \frac{m_2 M}{K m_1} \quad (52)$$

and the coefficients in Eq. 47 via (51) and (52)

$$D_{11} \simeq 2M/\kappa' \qquad D_{12} \simeq -m_1/\kappa' = -38M/\kappa'$$

$$D_{21} \simeq 2\frac{m_2 M}{m_1 \kappa'} = -0.05\ M/\kappa' \qquad D_{22} \simeq 2M/\kappa' \quad (53)$$

Arsenic (1) and Cu (2) in Ge fall within the scope of this example (2). These satisfy the above thermodynamic stability conditions (1) while indicating off-diagonal effects of the same magnitude as the on-diagonal. The sign of the interaction not surprisingly designates attraction between donors and attractors. If a diffusion couple could be constructed with m_1 uniform at 2×10^{-4} M and with a step function of 10^{-6} M in m_2 the profiles would evolve as in Fig. 2, curve 1. Were we to release the artificial symmetries introduced for conciseness such that $D_{11} >> D_{22}$, then the m_1 profile would be more like curve b and thus appear as more significant (cf. Ref. 1).

It may be observed that we chose a relatively large value of K to amplify the cross effect but such as to assure that the solution would remain stable. If m_2 were raised by a factor of slightly more than 2 then already det D_{ik} would change sign suggesting that precipitation of the ionic compound DA or one or the other of the solutes species occurs. This precipitation mode abetted by the jog in the diffusion penetration curve (Fig. 2) is a potential source of defects in semiconductors as pointed out by Hall and Racette (15). It is characteristic and not surprising that the magnitude of the diffusion cross-effect becomes significant only for compositions approaching the stability limit.

Complexing or pairing is not the only mechanism whereby strong diffusion cross-effects can appear in semiconductors. Electron-solute interactions can be equally strong leading to excessive and unwanted segregation in device structures Dorward (2) has demonstrated how highly mobile Cu segregates strongly to boron-rich regions in n-type silicon by this mechanism.

Discussion

It is evident that a rigorous treatment of all facets of diffusion in a doubly-doped semiconductor is extremely complex. Even assuming local equilibrium and the absence of significant correlation (13) and space charge effects there is a plethora of independent parameters such as equilibrium constants and diffusion rate constants which defy independent determination. It is probably for this reason that not a great deal has been added over the years to the pioneering efforts of Reiss et al (3). Notwithstanding, the present contribution comes close to defining the full scope of the problem and offers a firm framework for further experimentation and theoretical prediction.

References

1. J.S. Kirkaldy and D.J. Young, Diffusion in the Condensed State, pp. 119, 303, The Institute of Metals London (1987).
2. R.C. Dorward, Interactions Between Solute Atoms and Defects in Silicon and Germanium, pp. 145, 181, Ph.D. Dissertation, McMaster University (1967).
3. H. Reiss, C.S. Fuller and F.J. Morin, Bell Syst. Tech. J., 35, 535 (1956).
4. F.A. Kroger and H.J. Vink, Solid State Physics, 3, 307 (1956).
5. W. Schottky, in Halbeiterprobleme, Vol. 4, ed. W. Schottky, Braunschweig, Fr. Viewig (1958).
6. F.A. Kroger, The Chemistry of Imperfect Crystals, 2nd Ed., North-Holland, Amsterdam (1974).
7. D.J. Young and J.S. Kirkaldy, J. Phys. Chem. Solids, 45, 781 (1984).
8. D.J. Young and F. Gesmundo, Oxid. Met., 29, 169 (1988).
9. F. Seitz, Acta Cryst., 3, 355 (1950).
10. M.J. Dignam, D.J. Young and D.W.G. Goad, J. Phys. Chem. Solids, 34, 1227 (1973).
11. D.J. Young, E. DeLamotte and J.S. Kirkaldy, Can. J. Phys., 57, 722 (1979).
12. G.G. Libowitz, in Energetics in Metallurgical Phenomena, Vol. IV, ed. W.M. Mueller, p. 71, Gordon and Breach, New York (1968).
13. R.E. Howard and A.B. Lidiard, Reps. Prog. Phys, 27, 161 (1964).
14. C. Wagner, Thermodynamics of Alloys, Addison-Welsey, Reading, Mass (1952).
15. R.N. Hall and J.H. Racette, J. Appl. Phys., 35, 379 (1964).

Appendix

The following values were found for the kinetic coefficients l_i, λ_i;

$$l_1 = n\, m_D\, a_{V'}$$

$$l_2 = n \frac{m_h m_D}{m_h + m_D a_V x} a_V x$$

$$l_3 = n \frac{m_e m_D}{m_e + m_D a_V x} a_V x$$

$$l_4 = n \frac{m_D m_{V'}}{m_{V'} + m_D} a_V x$$

$$l_5 = n \frac{m_D m_e}{m_e + 2 m_D a_{V'}} a_{V'}$$

$$l_6 = n \frac{m_D m_h}{m_h + 2 m_D a_{V'}} a_{V'}$$

$$l_7 = n \frac{m_{Si} m_D m_e}{m_e m_D + m_e m_{Si} + 2 m_D m_{Si} a_V x} a_V x$$

$$l_8 = n \frac{m_{Si} m_D m_h}{m_h m_D + m_h m_{Si} + m_D m_{Si} a_V x} a_V x$$

$$l_9 = n \frac{m_{Si} m_{Si} m_D}{m_{Si} m_{Si} + 2 m_{Si} m_D} a_{V'}$$

$$l_{10} = n \frac{m_{Si} m_D m_e}{m_A m_e + m_D m_e + 2 m_D m_A a_V x} a_V x$$

$$l_{11} = n \frac{m_D m_A m_h}{m_A m_h + m_D m_h + 2 m_D m_A a_V x} a_V x$$

$$l_{12} = n \frac{m_D m_A}{m_D + m_A} a_V x$$

$$l_{13} = n \frac{m_D m_A m_{Si}}{m_A m_{Si} + m_{Si} m_D + 2 m_A m_D} a_V x$$

$$\lambda_1 = n\, m_A\, a_{V'}$$

$$\lambda_2 = n \frac{m_h m_A}{m_h + m_A a_V x} a_V x$$

$$\lambda_3 = n \frac{m_e m_A}{m_e + m_A a_V x} a_V x$$

$$\lambda_4 = n \frac{m_A m_{V'}}{m_{V'} + m_A} a_V x$$

$$\lambda_5 = n \frac{m_A m_e}{m_e + 2 m_A a_{V'}} a_{V'}$$

$$\lambda_6 = n \frac{m_A m_h}{m_h + 2 m_A a_{V'}} a_{V'}$$

$$\lambda_7 = n \frac{m_{Si} m_A m_e}{m_e m_A + m_e m_{Si} + 2 m_A m_{Si} a_V x} a_V x$$

$$\lambda_8 = n \frac{m_{Si} m_A m_h}{m_h m_D + m_h m_{Si} + m_D m_{Si} a_V x} a_V x$$

$$\lambda_9 = n \frac{m_{Si} m_{Si} m_A}{m_{Si} m_{Si} + 2 m_{Si} m_A} a_{V'}$$

$$\lambda_{10} = n \frac{m_A m_D m_e}{m_D m_e + m_A m_e + 2 m_A m_D a_V x} a_V x$$

$$\lambda_{11} = n \frac{m_A m_D m_h}{m_D m_h + m_A m_h + 2 m_D m_A a_V x} a_V x$$

$$\lambda_{12} = n \frac{m_A m_D}{m_A + m_D} a_V x$$

$$\lambda_{13} = n \frac{m_A m_D m_{Si}}{m_D m_{Si} + m_A m_{Si} + 2 m_D m_A} a_V x$$

In the above expressions $n = Nd^2/RT$ and, for conciseness, it has been assumed that

$$K_{iV}x\, \nu_{iV}x = K_{iV'}\, \nu_{iV'} = K_{iV}\, \nu_{iV}$$

Note that the m's here stand for primed quantities as in Eq. 18.

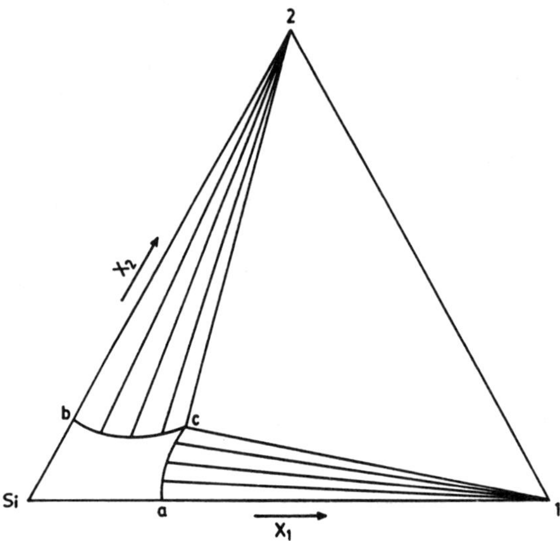

Fig. 1: Schematic phase diagram for donor (1) and acceptor (2) in silicon.

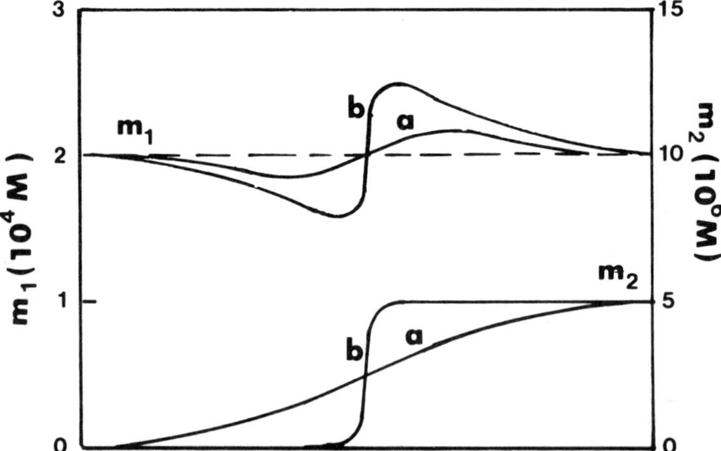

Fig. 2: Semi-quantitative representation of uphill diffusion of a donor (1) via electrostatic attraction to an acceptor (2) in a semi-conductor.

Atomistic approach of chemical diffusion phenomena

H. Sato and R. Kikuchi *
School of Materials Engineering, Purdue University,
West Lafayette, Indiana 47907, U.S.A.

Abstract

The application of the atomistic treatment of diffusion by means of the Cluster Variation method-Path Probability method formalism to chemical diffusion processes is briefly reviewed. The descriptions are intended to clarify relations among measurable quantities and underlying assumptions involved in traditional phenomenological approach.

* Permanent Address: Department of Materials Science and Engineering, University of California-Los Angeles, Los Angeles, California 90024.

1. Model of Diffusion

One of the major advantages of an atomistic treatment by means of the Cluster Variation method (CVM) and the Path Probability method (PPM) is that the Onsager equation for diffusion can be derived analytically with a few static and kinetic parameters. Therefore, the relations between measurable quantities even in complicated diffusion phenomena can be well understood.[1] Although intensely investigated, chemical diffusion phenomena in multicomponent systems are complicated, and misunderstandings can arise. The present treatment is therefore intended to clarify these relations and underlying assumptions involved in traditional phenomenological approach from the point of view of atomistic treatment.

Theoretical treatments of diffusion process in crystalline substances are generally based on a lattice gas model.[1,2] Here, the existence of an assembly of lattice sites is assumed. Atoms occupy these sites and can move through the lattice sites via vacant sites (vacancies). Generally, these lattice sites are spacially fixed. Such a model is called the rigid lattice gas model.

First, we assume a fundamental jump frequency of the form $w_i = \theta_i e^{-\frac{u_i}{kT}}$. for the ith species of constituent atoms (kinetic parameters). Because of mutual interactions among atoms, ε_{ij}, (we generally treat the case of nearest neighbor interactions only), an atom of the ith species jumping into the neighboring vacancy has to break the interactions with nearest neighboring atoms. This extra factor $B_i(\varepsilon_{ij})$, which makes the local jump frequency of an atom of the ith species in the form $\hat{w}_i = B_i(\varepsilon_{ij})\theta_i e^{-\frac{u_i}{kT}}$, is called the bond breaking factor.[1] The interactions, ε_{ij} (static parameters), determine the distribution of constituents in the crystal which is calculated by the CVM. The flow under this condition can then be calculated by the PPM.[1] The geometry of diffusion in alloys is indicated in Fig. 2.

2. Intrinsic Diffusion Coefficients, Tracer Diffusion Coefficients

The Onsager equations for diffusion

$$J_i = -\sum_{j=1}^{n} L_{ij}\frac{\partial \alpha_j}{\partial x}, \quad \alpha_i = \beta\mu_i \quad (\beta = 1/kT) \tag{1}$$

can thus be derived by the CVM-PPM formalism under the stationary state condition analytically.[1] Here, L_{ij} (the Onsager matrix) represents the mobility of particles under the equilibrium condition. As usual, J_i is the flow of particles, μ_i is the generalized chemical

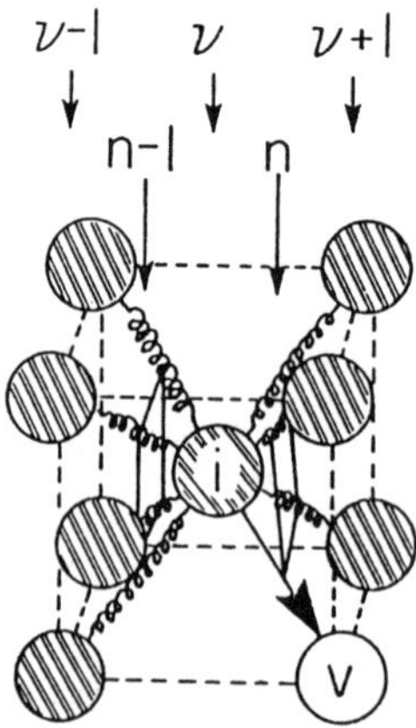

Fig. 1. Effect of bond breaking as the ith atom on the νth atomic plane jumps into a vacancy on the (ν+1)th plane. Springs indicate the bonds.

potential of the ith species, and x direction of flow. In order to keep the equilibrium distribution of particles during the stationary state diffusion process, appropriate sources and sinks of particles and vacancies may have to be attached to the system so that L_{ij} is a function of compositions only (with static and kinetic parameters). The reference system to be taken here is the laboratory frame fixed at a lattice plane. Equation (1) can be transformed to the form of the Fick's equation under the same condition, utilizing the relation

$$\frac{\partial \mu_i}{\partial x} = \sum_{j=1}^{n} \frac{\partial \mu_i}{\partial C_j} \frac{\partial C_j}{\partial x} \tag{2}$$

where C_i represents the composition of the ith species. Thus derived Fick's equation takes the form

$$J_i = -\sum_{j=1}^{n} D_{ij} \frac{\partial C_j}{\partial x} \tag{3}$$

Under the stationary state condition defined above, diffusion coefficients, D_{ij}'s, are the function of compositions only and do not depend on time or location. Let us call these diffusion coefficients the intrinsic diffusion coefficients.[1,3,4] The quantities D_{ij}'s are thus calculated analytically from the knowledge of L_{ij} which includes the jump frequencies $\hat{w}_i$'s. In this expression, J_i does not depend on the location of the specimen.

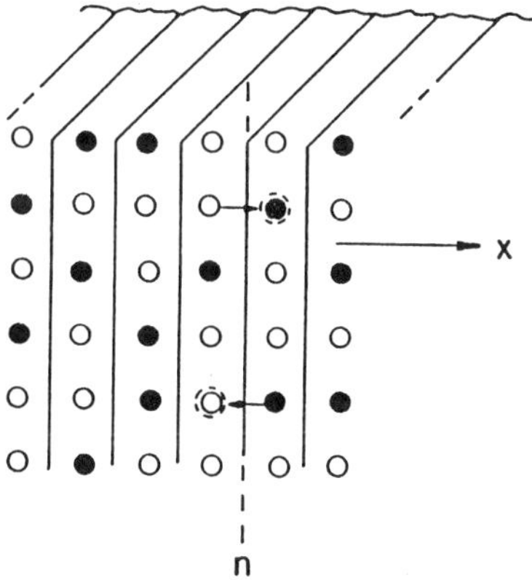

Fig. 2. Geometry of diffusion in alloys. Arrows indicate the motion of particles across the reference plane n perpendicular to the x-direction.

In an actual case, under the stationary state condition, D_{ij}'s cannot be measured by ordinary diffusion techniques. They can only be measured by methods which can measure the motion of particles under the equilibrium condition such as the inelastic neutron scattering. The tracer diffusion technique can also measure the motion of particles under equilibrium conditions, because the concentration of isotope atoms would be small enough and the concentration gradients of other species can be kept zero. In other words, the tracer diffusion technique measures the quantity directly related to the intrinsic diffusion coefficients.[1] Tracer diffusion coefficients can be calculated even at higher tracer concentrations by introducing the stationary state condition explicitly.[5,6,7] The calculation was made by the pair approximation of PPM.

3. Interdiffusion Coefficients

In an actual case in solids in the absence of sources and sinks defined above, diffusion under concentration gradients represents the interdiffusion (or the chemical diffusion). Because the diffusion coefficients D_{ij}'s are different from each other, the flow of materials in the +x direction and that in the -x direction are generally different. If the vacancy mechanism is assumed, the difference in the material flow, $\sum_i J_i$, in a closed system is compensated by the flow of vacancies J_v. In other words,

$$J_v = -\sum_i J_i \tag{3}$$

If this situation is kept uncorrected, the distribution of vacancies will change from the initial equilibrium distribution. Because, the mobility of particles depends on the vacancy concentration, this means that the diffusion coefficients would depend also on the location and time. This trouble can be avoided by an assumption that the distribution of vacancies in solids is kept at the equilibrium value while the diffusion process is taking place. This situation can generally be rationalized statistical-mechanically that the change of macrovariables such as the change of concentration in the diffusion process is much slower than the unit process such as the jumps of vacancies. If this condition is satisfied, diffusion coefficients do not change with time and location and the conversion process from the intrinsic diffusion coefficients defined earlier to interdiffusion coefficients in the Fick's equation for the closed system where the change of distribution of constituent species is allowed, can be made analytically.

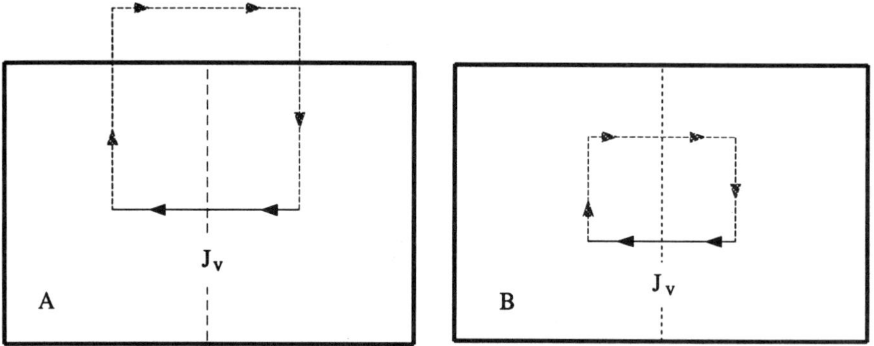

Fig. 3. Mechanism in regaining the equilibrium distribution of vacancies in the interdiffusion process.
A: Ideal soft lattice gas model.
B: Ideal rigid lattice gas model.

The process of regaining of the equilibrium distribution of vacancies in the chemical diffusion process does not need to be unique. Indeed, the analytical treatment can easily be made in the following two extreme cases. The one is what we call the ideal soft lattice gas model. As shown in Fig. 3, because of the flow of vacancies in one direction which compensates the material flow, one side of the specimen of a reference plane becomes vacancy rich, while the other side becomes vacancy deficient. Therefore, if the extra vacancies on one side are annihilated and the same number of vacancies are created on the other side, the equilibrium distribution is restored. Therefore, the path of vacancies during this restoring process is closed by going through the outside of the crystal as shown in Fig. 3A. Dislocations inside the crystal can be media for the annihilation and creation processes. Vacancies, based on the lattice gas model, are unoccupied sites. In other words, the annihilation and creation of vacancies correspond to the annihilation and the creation process of lattice sites. This is the origin of the name of the soft lattice as compared to the rigid lattice gas model where the number of the lattice sites is kept unchanged. On the side where the lattices are eliminated, a contraction in volume takes place while on the side where the lattice sites are created, the expansion takes place. As a result, the reference lattice plane moves towards the area where the vacancies are annihilated with the velocity v of vacancies across the reference lattice plane, if the volume change takes place only in the direction of diffusion. The velocity v is considered to be the Kirkendall velocity. In the ideal soft lattice gas model, the velocity v is thus calculated analytically based on the condition of interdiffusion (the system is closed)

$$\sum J_i + J_v = 0 \tag{4}$$

$$\sum C_i = 1 \tag{5}$$

Utilizing the expression for v, the Fick's equation for the open system (the expression corresponding to the Onsager equation) can be converted to the Fick's equation for the closed system as

$$J_i = -\sum_j \hat{D}_{ij} \frac{\partial C_j}{\partial x} \tag{6}$$

Here, the reference frame is the laboratory frame, and J_i represents the local flow across the reference plane fixed in the space. Between D_{ij} and $\hat{D}_{ij}$, the following relation exists:

$$\hat{D}_{ij} = D_{ij} - C_i \sum_{k=1}^{n} D_{ki} \tag{7}$$

Here, $\hat{D}_{ij}$'s are only a function of composition. Because the relation

$$\sum_{i=1}^{n} \hat{D}_{ij} = 0 \tag{8}$$

holds, the definition of n^2 interdiffusion coefficients in the above fashion is redundant, and we can define $(n-1)^2$ independent interdiffusion coefficients $\tilde{D}_{ij}$ as

$$\tilde{D}_{ij} = \hat{D}_{ij} - \hat{D}_{in} \tag{9}$$

Since $\hat{D}_{ij}$'s are calculated analytically, $\tilde{D}_{ij}$ can also be calculated analytically.[3,4]

The ideal soft lattice gas model is essentially the model adopted by Darken in his pioneering work of chemical diffusion.[8] However, the ideal soft lattice gas model is not the unique model to be taken. Because, vacancies are jumping back and forth rapidly during diffusion, equilibrium distribution of vacancies can also be achieved by this jumping back process.[9] In such a case we can call this model the ideal rigid lattice gas model. The schematic virtual vacancy path for the restoring of the equilibrium is then closed inside the solid like that shown in Fig. 3B. The motion of the reference lattice plane is then held back by the counter flow of vacancies and, hence, the lattice plane does not move. In other words, in this case, the

Kirkendall velocity v is zero and the Kirkendall effect does not exist although the chemical diffusion process is taking place.[9]

In the ideal rigid lattice gas model, the reference system is the laboratory frame and also is fixed at a lattice plane. Compared to the ideal soft lattice gas model, the diffusion coefficients describing the flow of particles across the lattice plane in this case are different due to the existence of the counter vacancy flow. It is essentially a flow in the laboratory frame fixed at a lattice plane. It is easily found that the interdiffusion coefficients based on the ideal rigid lattice gas model is exactly the same as those in the ideal soft lattice gas model with respect to the laboratory frame. In the ideal soft lattice gas model, because neither resultant material flow nor vacancy flow exists across the moving lattice plane, the diffusion coefficients measured at the moving lattice plane corresponds to the intrinsic diffusion coefficients. In Fig. 4, the relations between D_{ij}'s and $\hat{D}_{ij}$'s calculated for ideal solution and non ideal solution by the point approximation of the CVM (equivalent to the molecular field approach) are compared.

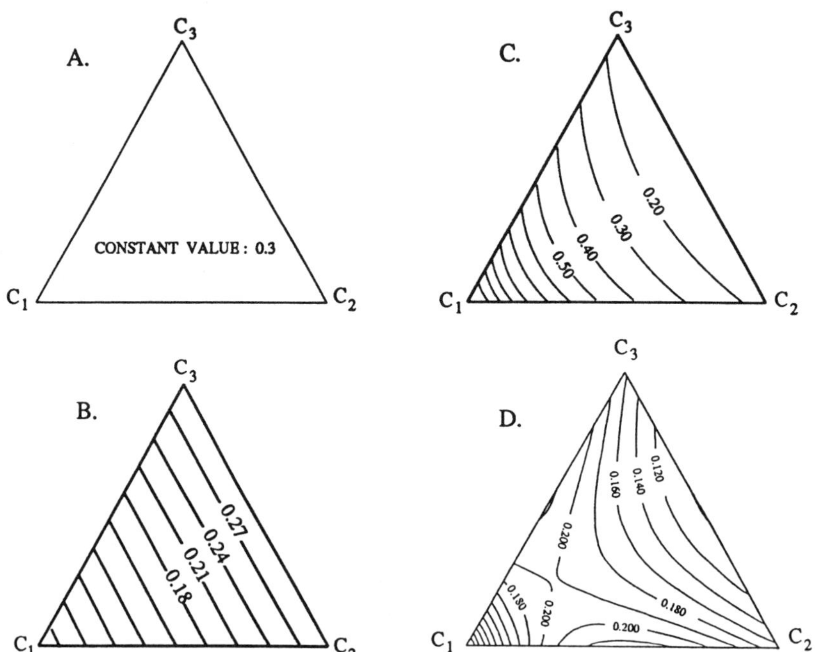

Fig. 4. Comparison of D_{ij}'s and $\hat{D}_{ij}$'s.

A. D_{11} for an ideal solid solution which corresponds to the case ε_{ij} = constant. For an ideal solution, D_{ij}'s are composition independent. Parameters are adjusted so that D_{11} takes a constant value 0.300.

B. $\hat{D}_{11}$ with the same parameters as A. It shows that $\hat{D}_{11}$ is linearly dependent on composition while corresponding D_{11} is composition independent.

C. D_{11} for a non ideal solid solution. It shows that D_{11} is composition dependent. Parameters ε_{ij}'s are arbitrarily taken.

D. $\hat{D}_{11}$ with the same parameters as D_{11} in C. It shows that $\hat{D}_{11}$ is far more strongly composition dependent than D_{11}.

4. The Ideal Soft Lattice Gas Model and the Ideal Rigid Lattice Gas Model in the Interdiffusion Process

Whether the soft lattice gas model or the rigid lattice gas model has to be used for the interdiffusion process depends on the type of materials. In metallic alloys, lattice sites are composed of constituent atoms. In such cases, soft lattice gas model is certainly proper. Observations of the Kirkendall effect in metallic alloys certainly justify this statement. On the other hand, in oxides, oxygen ions generally form a rigid lattice and provide lattice sites for metallic ions. The number of vacancies is determined by the oxygen partial pressure outside. In such a case, the creation and the annihilation of lattice sites are related to the behavior of oxygen ions and not to that of diffusing metallic ions. In such a case, the rigid lattice gas model should be more appropriate. In the limit of the ideal rigid lattice gas model, no Kirkendall effect is observed. In general, the situation would be in between. As long as the assumption that the distribution of vacancies is kept at the equilibrium value, the interdiffusion coefficients are uniquely determined from the knowledge of the intrinsic diffusion coefficients. The magnitude of the Kirkendall effect thus ranges between 0 and the maximum value associated with the case of the ideal soft lattice gas model.

Whether the assumption that vacancy distribution is always kept at the equilibrium holds or not is yet a problem to be confirmed. In the ideal soft lattice gas model, the problem depends on the source and the sink of vacancies such as dislocations. Bardeen and Herring once discussed this problem in detail and rationalized the assumption.[11] In the rigid lattice gas model, on the other hand, the problem depends on the ratio of the relaxation times by which the distribution of vacancies and of atoms reach the equilibrium distribution, respectively.[12] The ratio is roughly given by the concentration of vacancies,[12] and the lower is the vacancy concentration, the better the assumption holds. The vacancy concentration in metallic alloys is of the order of 10^{-4} to 10^{-5} while in oxides the ratio of 10^{-2} to 10^{-3} is common. Therefore, it seems that oxides have more difficulties in establishing the equilibrium distribution of vacancies than metallic alloys. Troubles by not satisfying the above condition is that the concentration distribution of constituents by chemical diffusion does not follow the Boltzmann's scaling law. Pfeifer, et al. measured the penetration curve of a diffusion couple consisting of FeO and MnO, and found that the penetration curves for different times deviated from the relation expected from that of $x/\sqrt{t}$.[13]

5. Diffusion Paths

Because it is possible to calculate the interdiffusion coefficients $\tilde{D}_{ij}$ analytically, it is possible to investigate the thermodynamical features of diffusion paths such as the prediction of the location of zero-flux planes based on certain thermodynamical data. Such examination was made on ternary systems.[4] For ternary systems, the diffusion path is obtained by the integration of two interdiffusion equations of the form

$$\begin{aligned} \frac{\partial C_1}{\partial t} &= \frac{\partial}{\partial x}\left[\tilde{D}_{11}\frac{\partial C_1}{\partial x} + \tilde{D}_{12}\frac{\partial C_2}{\partial x}\right] \\ \frac{\partial C_2}{\partial t} &= \frac{\partial}{\partial x}\left[\tilde{D}_{21}\frac{\partial C_1}{\partial x} + \tilde{D}_{21}\frac{\partial C_2}{\partial x}\right] \end{aligned} \tag{10}$$

which are derived analytically. Here, interdiffusion coefficients, $\tilde{D}_{ij}$'s, are calculated from the intrinsic diffusion coefficients analytically and, hence, these are functions of compositions only. Therefore, the diffusion path calculated should obey the Boltzmann relation and easy convergence to a stationary diffusion path was expected by the integration of interdiffusion equations by a finite difference method. The initial results indicated, in the case of ideal solutions in which D_{ij}'s are expected to be composition independent, the convergence was found to occur readily.[3] On the other hand, in cases where strong mutual interactions exist, the convergence became unreasonably slow.[3] In interdiffusion equations of multicomponent systems, the uniqueness of the solution cannot be proved rigorously unless $\tilde{D}_{ij}$'s are constant. Therefore, even the possibility of a solution which does not satisfy the Boltzmann's scaling law was suspected.[3] However, the difficulty of the convergence in the case of non-ideal solution, in which $\tilde{D}_{ij}$ is strongly composition dependent, was traced to a minor inappropriateness of the integration technique. This has cleared the suspicion.[14] This conclusion, however, does not guarantee that the Boltzmann scaling law should hold in experiments. It should be remembered that the assumption that the vacancy distribution is always kept under the equilibrium condition does not rigorously hold.[12,13]

Acknowledgement

The help of Mr. H. Zhang in preparing this manuscript is appreciated. The work was supported by DOE Grant DE-FG02-84ER45133.

References

1. H. Sato, "The Application of the Path Probability Method to Transport Phenomena," in Nontraditional Methods in Diffusion, edited by G. E. Murch, H. K. Birnbaum and J. R. Cost, TMS-AIME Conference Proceedings Volume (1984), pp. 203-235.
2. T. Ishikawa and H. Sato, "Phenomenological Theory for Superionic Transport," in Superionic Solids and Solid Electrolytes: Recent Trends, eds. A. L. Laskar and S. Chandra, Academic Press (1989), pp. 439-471.
3. T. Ishikawa, H. Sato, R. Kikuchi, S. A. Akbar, W. Zhu and A. Datta, "Chemical Diffusion in Multicomponent Alloys," to be published.
4. H. Sato, "Atomistic Treatment of Chemical Diffusion in Multicomponent Alloys: Theoretical Analysis of "Zero-Flux" Planes," in Diffusion Analysis and Applications, edited by A. D. Romig and M. A. Dayananda, TMS (1989), p. 19.
5. R. Kikuchi and H. Sato, J. Chem. Phys. *51,* 161 (1969).
6. R. Kikuchi and H. Sato, J. Chem. Phys. *53,* 2702 (1970).
7. R. Kikuchi and H. Sato, J. Chem. Phys. *57,* 4962 (1972).
8. L. S. Darken, Trans. AIME *174,* 184 (1948).
9. H. Sato and A. Datta, "Atomistic Treatment of Chemical Diffusion Phenomena in Oxides and in Metallic Alloys," TMS (1990).
10. H. Zhang, A. Datta and H. Sato, "Atomistic Treatment of Chemical Diffusion in Multicomponent Alloys: Relations between Intrinsic Diffusion Coefficients and Diffusion Coefficients in the Laboratory Frame," TMS (1990).
11. J. Bardeen and C. Herring, "Atom Movements," ASM, Cleveland (1951), p. 87.
12. T. Ishikawa, S. A. Akbar, W. Zhu and H. Sato, J. Amer. Cer. Soc., *71,* 513 (1988).
13. T. Pfeiffer, H. Schmalzried and Y. Ueshima, Ber. Bunsenges. Phys. Chem. *92,* 589 (1988).
14. H. Zhang, M.S. Thesis, School of Materials Engineering, Purdue University (May 1990).

Kirkendall and vacancy correlation effects in ternary diffusion

J.S. Kirkaldy
Institute for Materials Research, McMaster University, Hamilton, Ontario, Canada, L8S 4M1

Abstract

In ternary alloys there are two cross-effects in the 2×2 chemical D-matrix, one associated with the thermodynamic interactions and the other with the purely kinetic Kirkendall Effect. It is demonstrated by the development of explicit formulas that these effects contribute multiplicatively to the net diffusion within the diagonal coefficients and additively within the off-diagonal coefficients. The corrections due to vacancy correlation effects are evaluated according to a version of Manning's theory and are demonstrated to be practically significant to the diffusion profiles in special cases.

Introduction

Vacancy correlation effects in binary alloys are known to demand corrections to the tracer diffusion coefficients, particularly if the crystals are not close-packed (3-9). The equations due to Manning correcting Darken's Kirkendall velocity v and chemical coefficient $\tilde{D}$ are widely accepted even though the derivation possesses conceptual problems. At least five alternative constructions have been given (5-9) including our own (9). In the Manning (3) and Lidiard (8) treatments the assumptions are unnecessarily strong so the Onsager reciprocal relations are generated without the usual recourse to microscopic reversibility or detailed balance (9-11). Manning extended his calculation to ternary systems in 1970 (12) claiming to prove that correlation effects can have a large effect on interdiffusion cross-effects. However, he did not carry his calculations through to completion taking account of realistic cases and the actual laboratory frame solutions of the resulting ternary diffusion equations. It is one of our aims here to complete his calculation and thereby demonstrate that the error in neglecting vacancy correlation effects is not always of great practical significance with respect to the diffusion profiles.

Interpretation of the Manning results is based in part on two well-known results from ternary diffusion theory. Firstly, as regards the solutions of the Fick-type equations

$$\frac{\partial X_1}{\partial t} = \tilde{D}^3_{11} \frac{\partial^2 X_1}{\partial X^2} + \tilde{D}^3_{12} \frac{\partial^2 X_2}{\partial X^2} \tag{1}$$

and

$$\frac{\partial X_2}{\partial t} = \tilde{D}^3_{21} \frac{\partial^2 X_1}{\partial X^2} + \tilde{D}^3_{22} \frac{\partial^2 X_2}{\partial X^2} \tag{2}$$

if

$$\tilde{D}^3_{22} >> \tilde{D}^3_{11} \tag{3}$$

then the eigenvalues are $\tilde{D}3_{11}$ and $\tilde{D}3_{22}$ and the solutions are greatly simplified (13,14). Furthermore, in an experiment provided with countercurrent 1-3 constituents the gradient of the slow diffuser 1 has a strong effect on the fast diffuser 2 but not always the inverse. Secondly, the $\tilde{D}$-matrix for a ternary system is not unique since we are free to assign the solvent and kinetic order of the solutes arbitrarily. Kirkaldy et al. (15) presented the corresponding transformation formulas

$$\tilde{D}^n_{ik} = \tilde{D}^k_{nn} = \tilde{D}^n_{kk} = \tilde{D}^n_{ii} - \tilde{D}^n_{ii} - \tilde{D}^k_{ii} = -\tilde{D}^k_{im} \tag{4}$$

where the superscript makes clear the choice of solvent or dependent variable. The invariants of the matrix are its trace and determinant which determine the invariant profiles via the eigenvalues (10,15,16). The choice, while arbitrary, is by no means insignificant in the undertaking of a kinetic interpretation.

Manning's Formulas for Ideal Solutions

For a binary solution we have modified Manning's calculation to obtain via detailed balance the chemical coefficient in terms of tracer coefficients

$$\tilde{D} = \phi\,[X_b D^*_A + X_A D^*_B + X_A X_B (D^*_A - D^*_B)^2 (1 - \bar{f}) / \bar{f}\ \bar{D}\,] \tag{5}$$

where the average correlation factor (4) and tracer diffusion coefficient are

$$\bar{f} = X_A f_A + X_B f_B; \quad \bar{D} = X_A D^*_A + X_B D^*_B \tag{6}$$

and the thermodynamic factor is

$$\phi = 1 + \frac{d\,\ell n\,\gamma_A}{d\,\ell n\,X_A} = 1 + \frac{d\,\ell n\,\gamma_B}{d\,\ell n\,X_B} \tag{7}$$

Correspondingly, the Kirkendall velocity is

$$v = \frac{\phi}{\bar{f}} (D^*_A - D^*_B) \frac{\partial X_A}{\partial x}; \quad \partial X_A = -\,\partial X_B \tag{8}$$

The equations in the absence of correlation ($\bar{f}=1$) were first given by Darken (1). The formulas obtained invoking the unnecessary approximation $\bar{f}=f$, where f is the tracer correlation factor for the pure crystal (a so-called random alloy), are due to Manning (3,4).

Relations (5) and (6) can be subsumed within an Onsager L-matrix for the interdiffusion of the alloying elements and conserved vacancies in the Kirkendall or lattice frame of reference (4). This in our generalization of the Manning result can be trivially generalized from a 2×2 to an n×n matrix of the form (4, 12)

$$L_{ii} = \frac{X_i D_i^* \rho}{kT}\left[1 + \frac{2X_i D_i^*}{\bar{D}} \frac{(1-\bar{f})}{\bar{f}}\right] \tag{9}$$

and

$$L_{is} = \frac{2X_i X_s D_i^* D_s^* \rho}{kT\,\bar{D}} \frac{(1-\bar{f})}{\bar{f}} \tag{10}$$

where ρ is the density, with $\bar{f}$ and $\bar{D}$ generalized in the obvious way from Eq. 6. Note the necessary symmetry which follows from detailed balance, and the vanishing of the cross coefficients when correlation disappears ($\bar{f}=1$). The phenomenology for transformation to the chemical $\tilde{D}$-matrix in the laboratory frame was originally given by the author and coworkers (10,17,18). Using this procedure Manning has given explicit ($\bar{f}=f$) relations for obtaining the chemical $\tilde{D}$-matrix in the absence of thermodynamic effects (ideal solutions) but did not complete the calculation of the chemical $\tilde{D}$'s so as to assess the quantitative importance to the diffusion profiles of the off-diagonal correlation terms (cf. the trailing term in Eq. 5). We now undertake to do this. Manning's ideal solution relations for an n-component system corrected with $f \rightarrow \bar{f}$ are given in Kirkaldy et al.'s notation (15) by

$$\tilde{D}_{ik}^{n} = D_{ij} - D_{in} \tag{11}$$

where (12)

$$D_{ii} = D_i^*\{1 - X_i[1 + (1 - D_i^*/\bar{D})(1-\bar{f})/\bar{f}]\} \tag{12}$$

and

$$D_{is} = -D_s^* X_i\{1 + (1 - D_i^*/\bar{D})(1-\bar{f})/\bar{f}]\};\; s \neq i \tag{13}$$

For example with n = 3

$$\tilde{D}_{11}^{3} = D_1^*(X_2+X_3) + D_3^* X_1 + X_1 \frac{(D_3 - D_1^*)(\bar{D} - D_1^*)(1-\bar{f})}{\bar{D}\,\bar{f}} \tag{14}$$

and

$$\tilde{D}_{12}^{3} = (D_3^* - D_2^*) X_1 \left\{\frac{\bar{D} - (1-\bar{f}) D_1^*}{\bar{D}\,\bar{f}}\right\} \tag{15}$$

Clearly Eq. 14 for the on-diagonal coefficient reduces to the binary case when n = 2 (3→2) and indeed the formula appears roughly as for a pseudo-binary with 2+3 as a single component. Since the correction remains quadratic in the mole fractions we may expect it to be of the order of 10% for an f.c.c. crystal ($f \simeq 0.78$) and somewhat higher for b.c.c. ($f \simeq 0.72$), differences which are not easily detectable in a solid state experiment. Eq. 15 demonstrates that as earlier pointed out by Lane and Kirkaldy (17,18) the main cross-interaction is via the Kirkendall velocity itself (cf. Eq. 8), rather than via the associated correlation effects. That is to say, the effect on component 1 is due to the pseudo-binary Kirkendall effect associated with interdiffusion of components 2 and 3 as indexed by a drift velocity roughly proportional to $(D^*_3 - D^*_2)$ (cf. Eq. 8). To assess the relative magnitudes it is convenient to choose 2 as the fastest diffuser and 3 as the slowest diffuser to maximize the pseudo-binary Kirkendall effect thus specifying D_1^* as approaching D and vanishing of the correlation effect. In a diffusion couple with 1 initially uniform and 2 and 3 with countercurrent step functions the Kirkendall flow will be towards the 2-rich regions and this will carry 1 in the same direction simulating uphill diffusion of 1 on its own gradient and on that of 2 ($D^*_3 - D^*_1 < 0$). In the symmetric ideal solution experiment with X_2 initially uniform one may expect an uphill effect symmetric in magnitude since the higher 2 mobility effect will be moderated by the reduced Kirkendall velocity generated by 1 and 3. This is in contrast to thermodynamic interactions where the separation forces are symmetric but the relative mobility effect is not (see below).

Effect of Combined Thermodynamics and Kirkendall Motion

If with Lane and Kirkaldy (17,18,10) we neglect correlation effects and concentrate on near terminal solutions in the Gibbs triangle then we can write down general formulas. Here the solution thermodynamics of the independent species is represented in terms of Wagner Taylor expansions for activity coefficients of the form

$$\ell n\,\gamma_1 = \ell n\,\gamma_1^0 + \varepsilon_{11} X_1 + \varepsilon_{12} X_2 \tag{16}$$

and

$$\ell n\,\gamma_2 = \ell n\,\gamma_2^0 + \varepsilon_{21} X_1 + \varepsilon_{22} X_2 \tag{17}$$

where the ε matrix is symmetric. Corresponding to Eqs. 14 and 15 we have from Ref. 10

$$\tilde{D}^3_{11} = \left\{D^*_1(X_2 + X_3) + D^*_3 X_1\right\}(1 + \varepsilon_{11} X_1) + \varepsilon_{12} X_1 X_2 (D^*_3 - D^*_2) \tag{18}$$

and

$$\tilde{D}^3_{12} = X_1\left\{(D^*_3 - D^*_2) + \left[D^*_1(X_2 + X_3) + D^*_3(2X_2 - X_1)\right]\varepsilon_{12} + \left[D^*_3(2\varepsilon_{11} X_1 - \varepsilon_{22} X_2) - D^*_2\varepsilon_{22} X_2\right]\right\}$$

$$\simeq X_1\left\{(D^*_3 - D^*_2) + \varepsilon_{12} D^*_1\right\} \quad ; \quad X_3 \simeq 1 >> X_1, X_2 \tag{19}$$

Noting the diluteness we can infer that the trace and determinant of the matrix are positive definite indicating stability against spinodal decomposition. Furthermore we can indicate the relative magnitude of cross-effects via the approximations

$$\frac{\tilde{D}^3_{12}}{\tilde{D}^3_{11}} \simeq \frac{D^*_3 - D^*_2}{D^*_1} X_1 + \varepsilon_{12} X_1 \quad ; \quad \frac{\tilde{D}^3_{21}}{\tilde{D}^3_{22}} \simeq \frac{D^*_3 - D^*_1}{D^*_2} X_2 + \varepsilon_{12} X_2 \tag{20}$$

which clearly indicates how in a symmetric pair of experiments with $D^*_2 > D^*_1 > D^*_3$ and $\varepsilon_{12} = 0$ the profile cross-effect also tends to symmetry (cf. discussion of Eqs. 1-3) which is not the case for the kinetically independent thermodynamic effect (see below).

Clearly, the kinetic and thermodynamic cross-effects contribute additively to the off-diagonal coefficients and are generally competitive in magnitude ($|\varepsilon_{12}| \sim 4 - 20$). On the other hand, they contribute multiplicatively to the on-diagonal correction term (Eq. 18). Since ε_{12} can be quite large this term can be significant if the Kirkendall velocity, proportional to $D^*_3 - D^*_2$, is also large. This is a previously unrecognized contribution to ternary diffusion interactions.

Discussion

The two controlling extremes corresponding to Eq. 20 are well-known. For example, in Fe-Si-C (3 2 1) $\varepsilon_{12} \simeq 9.5$ and $|D^*_3 - D^*_2| << D^*_1$, and these through solutions (1) and (2) quantitatively generate Darken's famous experimental results for uphill diffusion of C by thermodynamic action of a Si gradient (13, 19). Note that the symmetric experiment driven by a C gradient would fail to displace Si. The corresponding experiment by Sabatier and Vignes (20, 21) in Fe-Co-Ni (3-2-1) which is nearly ideal, has been explained satisfactorily in terms of the relative tracer and therefore the Kirkendall motion (21, 10).

While one can construct special conditions where vacancy correlation effects as measured by 1-f becomes theoretically significant the current accuracy of diffusion measurements in solids precludes direct observation through the chemical coefficients. However, as in the binary system such effects appear more strongly in the Kirkendall velocity (4, 9, 10), a matter which deserves further exploration.

In binary liquid diffusion the diffusion coefficient measurements are sufficiently accurate to establish significant deviations from the Darken predictions (Table 1; 22, 10). This correlation effect, which has the same positive sign as for solids, is not understood insofar as we can ascertain.

Table 1

Deviations between observed and predicted values of $\tilde{D}$ based on the D-H-C equation. After Tyrrell and Harris(22) $\Delta = 100[1 - (X_1D^*_2 + X_2D^*_1)\phi/D]$

System	Δ
Benzene-chlorobenzene	0.7
Cyclopentane-cyclooctane	3.9
n-Dodecane-n-octane	3.2
n-Dodecane-n-hexane	2.9
Benzene-n hexane	12.1
Benzene-n heptane	11.7
Benzene-cyclohexane	9.5
Benzene-OMCTS[a]	9.0
Tetrachloromethane-OMCTS	4.3

[a] The intradiffusion coefficient of octamethyl cyclotetrasiloxane (OMCTS) used was obtained for a mixture with perdeuteriobenzene.

References

1. L.S. Darken, Trans. AIME, 175, 184 (1948).
2. J. Bardeen and C. Herring, Atom Movements, p. 87, ASM, Cleveland (1951).
3. J.R. Manning, Acta Met., 15, 847 (1967).
4. J.R. Manning, Diffusion Kinetics for Atoms in Crystal, D. Van Nostrand Inc., Princeton, N.J., (1968).

5. K.P. Gurov, A.V. Nazarov and K.K. Anandybov, Fiz. Metal. Metalloved, 32, 176 (1971).
6. Th. Heumann, J. Phys. F. 9, 1997 (1979).
7. M.A. Dayananda, Acta Met., 29, 1151 (1981).
8. A.B. Lidiard, Acta Met., submitted 1986.
9. J.S. Kirkaldy, Scripta Met. 21, 33 (1987).
10. J.S. Kirkaldy and D.J. Young, Diffusion in the Condensed State, Institute of Metals, London (1987).
11. J.S. Kirkaldy, D.J. Young and J.E. Lane, Acta Met., 35, 1273 (1987).
12. J.R. Manning, Met. Trans., 1, 499 (1970).
13. J.S. Kirkaldy, Can. J. Phys. 35, 435 (1957).
14. J.S. Kirkaldy, Can. J. Phys. 35, 809 (1958).
15. J.S. Kirkaldy, J.E. Lane and G.R. Mason, Can. J. Phys., 41, 2174 (1963).
16. T.O. Ziebold and R.E. Ogilvie, Trans. AIME, 239, 942 (1967).
17. J.E. Lane and J.S. Kirkaldy, Can. J. Phys. 42, 1643 (1964).
18 J.S. Kirkaldy and J.E. Lane, Can. J. Phys., 44, 2059 (1966).
19. L.S. Darken, Trans. AIME, 180, 430 (1949).
20. J.P. Sabatier and A. Vignes, Mém. Sci. Rev. Mét., 64, 225 (1967).
21. A. Vignes and J.P. Sabatier, Trans. AIME, 245, 1795 (1969).
22. H.J.V. Tyrrell and K.R. Harris, Diffusion in Liquids, Butterworth & Co. Ltd (1984).

Equal eigenvalues in multicomponent diffusion: the extraction of diffusion coefficients from experimental data in ternary systems*

D.G. Miller

Chemistry and Materials Science, Lawrence Livermore National Laboratory, Livermore, California 94550, U.S.A.

ABSTRACT

There has been considerable theoretical and experimental work on ternary systems whose diffusion coefficient matrix has two distinct eigenvalues. This is the most common case. However, with the increasing number of ternary diffusion measurements in liquids, cases have arisen where the eigenvalues may be equal and where the usual equations do not apply.

Little work has been done on the solution of the equal eigenvalues case, and none has been done on the connection of this solution to experimental data. This paper will present new explicit expressions connecting ternary diffusion coefficients to experimental data for the precise Gouy and Rayleigh optical methods. Some new results for the two 4-component cases will also be presented.

Introduction

There has been considerable theoretical work on ternary diffusion where the two eigenvalues of the diffusion coefficient matrix are distinct, beginning with Fujita and Gosting(1) and Kirkaldy(2). General methods for multicomponent systems were provided by Kirkaldy(3) and Toor(4). Explicit results for 4-component systems were given by Toor(4), and in more detail by Kim(5). Methods for extraction of diffusion coefficients from experimental data for the precise Gouy and Rayleigh optical methods are available, (3-component Gouy(6); 3- and 4-component Rayleigh and Gouy(7)). However, the increasing number of diffusion measurements on ternary liquid systems has led to cases where the eigenvalues may be equal, so that the usual equations do not apply.

Little theoretical work has been done on the equal eigenvalue case for ternary systems and none for 4-component systems. The solution for a special ternary case was first discussed by Kirkaldy(2), and then by Sundelof and Sodervi(8). The general ternary solution was given by Toor(4). No connection to experiment exists for either 3- or 4-component systems. The former has one case, where the two eigenvalues are equal. The latter has two cases, one where two of the three eigenvalues are equal, and the other where all three are equal.

This paper will present the known solution of the diffusion equation for the ternary equal eigenvalue case and present new explicit expressions for the diffusion coefficients in terms of experimentally derived quantities for the Gouy and Rayleigh optical methods, specialized to "free diffusion" boundary conditions. Some new results for the two 4-component cases will be presented. L'Hôpital's Rule was used by Kirkaldy; it is also used here.

General Equations for Distinct Eigenvalues

The fastest way to reach our results is through the matrix method of Toor(4). Toor's method requires that the diffusion coefficients be concentration independent, which is a good assumption if experimental concentration differences are small. The method also depends on constant initial and boundary conditions for the diffusion equations. This condition is satisfied by most of the usual experimental methods. The general idea of Toor's method is to transform the

*Work performed under the auspices of the U.S. Department of Energy by the Lawrence Livermore National Laboratory under contract No. W-7405-ENG-48.

coupled diffusion equations to an equivalent set of uncoupled binary diffusion equations, each of which involves a "combine concentration" and a "combine diffusion coefficient" equal to the corresponding eigenvalue of the diffusion coefficient matrix. Each binary equation is solved in terms of its combine concentration, and the set of solutions is transformed back to solutions in terms of the original stoichiometric concentrations.

We shall give Toor's results in a slightly different notation. Let **D** be diffusion coefficient matrix (components D_{ij}) on the *volume-fixed reference frame* (9). Let **c** be the column vector of the original stoichiometric concentrations (components c_i). Let **f** be the matrix of transformed solutions of the binary case (components f_{ij}), which involve the binary solutions F_k in terms of the k th eigenvalue λ_k.

The general solution of the diffusion equations has the form

$$\delta \mathbf{c} = \mathbf{f} \cdot \delta \mathbf{c}^{\mathbf{o}} \tag{1}$$

where $\delta\mathbf{c}$ (components δc_i) reflect c_i as a function of distance x and time t, and $\delta\mathbf{c}^{\mathbf{o}}$ (components $\delta c_i{}^o$) reflect the initial conditions. Their expressions and F_k depend on the initial and boundary conditions.

Toor has shown(4) that

$$\mathbf{f} = \sum_k F_k \mathbf{Z}_k \tag{2}$$

where $\mathbf{Z}_k$ (components $(Z_{ij})_k$) for the *distinct eigenvalues case* is in turn given by Sylvester's theorem(4)

$$\mathbf{Z}_k = \frac{\prod_{m \neq k} (\lambda_m \mathbf{I} - \mathbf{D})}{\prod_{m \neq k} (\lambda_m - \lambda_k)} \tag{3}$$

where **I** is the unit matrix. Explicit expressions for $(Z_{ij})_k$ will be found by multiplying out the matrices and putting the result in component form. Note that $\mathbf{Z}_k$ is independent of the initial and boundary conditions, i.e. is the same for any set of such conditions(4,5), whereas F_k, $\delta\mathbf{c}$, and $\delta\mathbf{c}^{\mathbf{o}}$ do depend on those conditions.

Given Eqs. 1-3, the component form of f_{ij} becomes

$$f_{ij} = \sum_k (Z_{ij})_k F_k \tag{4}$$

and the component form of the concentrations becomes

$$\delta c_i = \sum_j f_{ij} \delta c_j^o = \sum_j \sum_k F_k (Z_{ij})_k \, \delta c_j^o \tag{5}$$

We shall be primarily interested in the free diffusion initial and boundary conditions. These are as follows. Consider two infinitely long uniform columns of solution whose concentrations are slightly different. These columns are placed with the least dense on top of the more dense, and are initially separated by an infinitely sharp boundary at distance x = 0. At time t = 0, the solutions are allowed to diffuse into each other. In this circumstance

$$\delta c_i = c_i - \bar{c}_i \tag{6}$$

$$\delta c_i^o = \Delta c_i/2 \tag{7}$$

$$\bar{c}_i = (c_{i\,\text{bottom}} + c_{i\,\text{top}})/2 \tag{8}$$

$$\Delta c_i = (c_{i\,\text{bottom}} - c_{i\,\text{top}}) \tag{9}$$

$$F_k = \text{erf}(y/\lambda_k^{1/2}) \tag{10}$$

$$y = x/(2t^{1/2}) \tag{11}$$

Let us now look at the refractive index. If concentration differences are small, the refractive index n can be written in linear form,

$$n - \bar{n} = \sum R_i(c_i - \bar{c}_i) \tag{12}$$

where R_i is the refractive index increment of constituent i, $\partial n_i/\partial c_i$. Therefore, Eqs. 4, 5, 6, and 7 yield

$$n - \bar{n} = \sum_i \sum_j R_i f_{ij} \frac{\Delta c_j}{2} = \sum_i \sum_j \sum_k R_i (Z_{ij})_k \frac{\Delta c_j}{2} F_k \tag{13}$$

It is convenient to define the refractive index fraction α_i for the free diffusion case:

$$\alpha_i = R_i \Delta c_i / \Delta n \tag{14}$$

where

$$\Delta n = n_{\text{bottom}} - n_{\text{top}} \tag{15}$$

and

$$\Sigma \alpha_i = 1 \tag{16}$$

We can now rewrite Eq. 13 as

$$n - \bar{n} = \frac{\Delta n}{2} \sum_i \sum_j \sum_k (Z_{ij})_k \, F_k \, \frac{R_i}{R_j} \, \alpha_j \tag{17}$$

Let us now define a quantity Γ_k

$$\Gamma_k = \sum_i \sum_j \frac{R_i}{R_j} (Z_{ij})_k \, \alpha_j \tag{18}$$

Then Eq. 17 becomes

$$n - \bar{n} = \frac{\Delta n}{2} \sum_k \Gamma_k F_k \qquad (19)$$

It can be shown that

$$\sum_k \Gamma_k = 1 \qquad (20)$$

The interferometric fringes of the Rayleigh and Gouy optical methods results from the variation of refractive index in the diffusion boundary(10). We restrict ourselves to the free diffusion case.

Rayleigh interferometry defines a reduced fringe number f(j) which is given by

$$f(j) = \frac{2j - J}{J} = \frac{n - \bar{n}}{(\Delta n/2)} \qquad (21)$$

where j is the number of the fringe whose corresponding position in the diffusion cell is x_j, and J is the total number of fringes (proportional to Δn). Consequently from Eq. 19,

$$f(j) = \sum_k \Gamma_k F_k \qquad (22)$$

where F_k is given by Eq. 10.

For Gouy interferometry in the free diffusion case, it was shown(1,6) that the Gouy fringe positions Y_j are proportional to dn/dy and that the interference condition is proportional to $[(n-\bar{n}) - y\,dn/dy]$. For the distinct eigenvalue case, the result for the reduced fringe position Y_j^* (defined by Eq. 23) is

$$Y_j^* = Y_j t^{1/2} = K \sum_k \Gamma_k (1/\lambda_k^{1/2}) \exp(-y_j^2/\lambda_k) \qquad (23)$$

where K is a function of apparatus constants(7) and y_j is defined by the interference condition for fringe number j. This interference condition is(6)

$$\frac{Z_j}{J} = \sum_k \Gamma_k \left[F_k - y \frac{\partial F_k}{\partial y} \right] = \sum_k \Gamma_k f(y_j/\lambda_k^{1/2}) \qquad (24)$$

where

$$Z_j \cong j + 3/4 \qquad (25)$$

for a dark fringe(11,12), F_k is the error function of Eq. 10, and the function f(z) is

$$f(z) = \operatorname{erf}(z) - \frac{2z \exp(-z^2)}{\pi^{1/2}} \qquad (26)$$

Let us now evaluate $(Z_{ij})_k$ for the distinct eigenvalue case. For a ternary system, Eq. 3 yields

$$\mathbf{Z}_k = \frac{\lambda_m \mathbf{I} - \mathbf{D}}{(\lambda_m - \lambda_k)} \qquad m \neq k \tag{27}$$

and in component form

$$(Z_{ij})_k = \frac{\delta_{ij}\lambda_m - D_{ij}}{\lambda_m - \lambda_k} \qquad m \neq k \tag{28}$$

where δ_{ij} is the Kronecker delta. For a quaternary system, Eq. 3 yields

$$\mathbf{Z}_k = \frac{\lambda_m\lambda_n\mathbf{I} - (\lambda_m + \lambda_n)\mathbf{D} + \mathbf{D}\cdot\mathbf{D}}{(\lambda_m - \lambda_k)(\lambda_n - \lambda_k)} \qquad m \neq n \neq k \tag{29}$$

and in component form

$$(Z_{ij})_k = \frac{\delta_{ij}\lambda_m\lambda_n - (\lambda_m + \lambda_n)D_{ij} + \sum_l D_{il}D_{lj}}{(\lambda_m - \lambda_k)\,(\lambda_n - \lambda_k)} \qquad m \neq n \neq k \tag{30}$$

Ternary Systems

We now specialize our equations to ternary systems. From Eqs. 5 and 28, we obtain

$$c_i - \bar{c}_i = \sum_j \left[\frac{(\lambda_2\delta_{ij} - D_{ij})F_1 - (\lambda_1\delta_{ij} - D_{ij})F_2}{\lambda_2 - \lambda_1} \right] \frac{\Delta c_j}{2} \tag{31}$$

From Eqs. 18 and 28 we obtain more specifically

$$\Gamma_1 = \frac{(X_1 - \lambda_2)\alpha_1 + (X_2 - \lambda_2)\alpha_2}{(\lambda_1 - \lambda_2)} = \frac{X_1\alpha_1 + X_2\alpha_2 - \lambda_2}{(\lambda_1 - \lambda_2)} \tag{32}$$

$$\Gamma_2 = 1 - \Gamma_1 = \frac{-[X_1\alpha_1 + X_2\alpha_2 - \lambda_1]}{\lambda_1 - \lambda_2} \tag{33}$$

where

$$X_1 = D_{11} + \frac{R_2}{R_1} D_{21} \tag{34}$$

$$X_2 = D_{22} + \frac{R_1}{R_2} D_{12} \tag{35}$$

Therefore, we can write Eq. 19 as

$$n - \bar{n} = \frac{\Delta n}{2}\left[\frac{(X_1\alpha_1 + X_2\alpha_2 - \lambda_2)F_1 - (X_1\alpha_1 + X_2\alpha_2 - \lambda_1)F_2}{\lambda_1 - \lambda_2} \right] \tag{36}$$

A. Rayleigh Interferometry

From Eq. 21 and Eq. 36, we obtain an expression for the Rayleigh f(j) which is suitable for non-linear least-squares in the distinct eigenvalues case.

$$f(j) = \frac{(X_1\alpha_1 + X_2\alpha_2 - \lambda_2)\,\mathrm{erf}(y_j/\lambda_1^{1/2}) - (X_1\alpha_1 + X_2\alpha_2 - \lambda_1)\,\mathrm{erf}(y_j/\lambda_2^{1/2})}{\lambda_1 - \lambda_2} \tag{37}$$

Notice that there are four quantities X_1, X_2, λ_1, and λ_2 in Eq. 37, which are functions of the four D_{ij} and two R_i. The experimental quantities are J, j, x_j, and t. For a single experiment, it is possible to least-square fringe position data to get $X_1\alpha_1 + X_2\alpha_2$, λ_1, and λ_2. However, to get X_1 and X_2 separately requires at least two experiments with different values of α_1, ($\alpha_2 = 1-\alpha_1$), i.e., different $\Delta c_1/(\Delta c_2)$ ratios. The R_i are obtained from at least two experiments with different $\Delta c_1/\Delta c_2$ using Eqs. 14 and 15, i.e.,

$$\Delta n = R_1\Delta c_1 + R_2\Delta c_2 \tag{38}$$

It can be shown that the four D_{ij} are related to the X_1, X_2, λ_1, λ_2, and R_1, R_2 by

$$D_{ij} = (-1)^i \frac{R_j}{R_i}\left[\frac{T_2 + X_j\,(X_1 + X_2 - X_i - T_1)}{X_1 - X_2}\right] \tag{39}$$

where for a 2x2 matrix, T_1 and T_2 are the traces of first and second order, respectively(13)

$$T_1 = \lambda_1 + \lambda_2 = D_{11} + D_{22} \tag{40}$$

$$T_2 = \lambda_1\lambda_2 = \det(\mathbf{D}) = D_{11}D_{22} - D_{12}D_{21} \tag{41}$$

These results were originally written in other forms(7,14), which are convenient for the distinct eigenvalue case. For example, Γ_1 can also be written as

$$\Gamma_1 = a + b\alpha_1 \tag{42}$$

where

$$a = \frac{X_1 - \lambda_2}{\lambda_1 - \lambda_2} \tag{43}$$

$$b = \frac{X_1 - X_2}{\lambda_1 - \lambda_2} \tag{44}$$

These expressions simplify the least-squares calculations, but are less suitable for exploring the equal eigenvalue case.

Let us now look at the equal eigenvalue case, where $\lambda_1 = \lambda_2$. Clearly Eqs. 31, 36, and 37 become 0/0. We can find their limiting values by means of L'Hôpital's' rule, differentiating numerator and denominator with respect to λ_2, and then setting $\lambda_2 = \lambda_1$.

The result for $c_i - \bar{c}_i$ is(4)

$$c_1 - \bar{c}_1 = \frac{\Delta c_1}{2} F_1 + \left[(D_{11} - \lambda_1)\frac{\Delta c_1}{2} + D_{12}\frac{\Delta c_2}{2}\right]\frac{\partial F_1}{\partial \lambda_1} \tag{45}$$

$$c_2 - \bar{c}_2 = \frac{\Delta c_2}{2} F_1 + \left[(D_{22} - \lambda_1)\frac{\Delta c_2}{2} + D_{21}\frac{\Delta c_1}{2}\right]\frac{\partial F_1}{\partial \lambda_1} \tag{46}$$

where F_1 is given by Eq. 10, and

$$\frac{\partial F_1}{\partial \lambda_1} = \frac{-y_j \exp(-y_j^2/\lambda_1)}{\pi^{1/2}\lambda_1^{3/2}} \tag{47}$$

The expression for f(j) (or equivalently $(n-\bar{n})/(\Delta n/2)$) can be found either by substitution of $c_i - \bar{c}_i$ from Eq. 45 and 46 into Eq. 12, or by applying L'Hôpital's rule directly to Eq. 37. This expression is

$$f(j) = \operatorname{erf}(y_j/\lambda_1^{1/2}) - \frac{[X_1\alpha_1 + X_2\alpha_2 - \lambda_1]\, y_j \exp(-y_j^2/\lambda_1)}{\pi^{1/2}\lambda_1^{3/2}} \tag{48}$$

where analogs of the Γ_i can be defined as

$$\Gamma_{1e} = 1 \tag{49}$$

$$\Gamma_{2e} = X_1\alpha_1 + X_2\alpha_2 - \lambda_1 \tag{50}$$

We note that Γ_{2e} is the negative of the numerator of the Γ_2 of the distinct eigenvalue case (Eq. 33). The least-square variables here are X_1, X_2, and λ_1, which together with the condition $\lambda_1 = \lambda_2$, yield a solution for the D_{ij}. That solution is identical with Eq. 39, with T_1 and T_2 replaced by their limiting values

$$T_1 = 2\lambda_1 \tag{51}$$

$$T_2 = \lambda_1^2 \tag{52}$$

B. Gouy Interferometry

Consider the distinct eigenvalue case. The fringe position expression (Eq. 23) and the interference condition (Eq. 24) for Gouy interferometry have the same expressions for Γ_1 and Γ_2 as does the Rayleigh f(j) (Eq. 37). Therefore the determination of the D_{ij} is related to the Rayleigh procedure for the distinct eigenvalue case(7). However, both Eq. 23 and Eq. 24 are needed because j, J, Y_j, and t are the experimental variables, whereas y_j is not. Consequently evaluating the least-square parameters X_1, X_2, λ_1, and λ_2 requires an iterative least-squares calculation involving both equations, as described elsewhere(7). Since these least-squares parameters are exactly the same as in the Rayleigh equations, the solution for D_{ij} (Eq. 39) is exactly the same.

Now let us examine the case where $\lambda_1 = \lambda_2$. L'Hôpital's rule applied to the expressions for Y_j^* and Z_j/J for the distinct case leads to the results:

$$Y_j^* = K\frac{[2\lambda_1^2 + \Gamma_{2e}(2y_j^2 - \lambda_1)]\exp(-y_j^2/\lambda_1)}{2\lambda_1^{5/2}} \tag{53}$$

and

$$\frac{Z_j}{J} = f(y_j/\lambda_1^{1/2}) - \frac{2\Gamma_{2e} y_j^3 \exp(-y_j^2/\lambda_1)}{\pi^{1/2}\lambda_1^{5/2}} \tag{54}$$

where Γ_{2e} is given in terms of X_1, X_2, and λ_1 by Eq. 50. Just as for the distinct case, experimental variables are j, J, and Y_j, and t. The least-square variables are X_1, X_2, and λ_1. The values of y_j to use in Eq. 53 are obtained by iteration from Eq. 54. Actually, a more complex, iterative nonlinear least-squares procedure must be used, which is analogous to that used in the distinct case (7). The Γ_{2e}, and its corresponding X_1, X_2, and λ_1, are exactly the same as in the equal eigenvalue Rayleigh case, so that the solution for the D_{ij} in terms of X_1, X_2, R_1, R_2, and λ_1 is also exactly the same.

We emphasize that the least-square variables for both Rayleigh and Gouy, as expressed in Γ_k or Γ_{2e}, are the same in both the distinct and equal eigenvalues cases, only their multiplying functions are different.

Quaternary Systems

For brevity, we shall limit our discussion to the Rayleigh case, which is the simplest algebraically. The analysis is analogous to the ternary system case.

A. Distinct Eigenvalues

For this case(4),

$$F_k = \operatorname{erf}(y_j/\lambda_k^{1/2}) \tag{55}$$

and Eq. 30 yields

$$(Z_{ij})_1 = \frac{\delta_{ij}\lambda_2\lambda_3 - (\lambda_2 + \lambda_3)D_{ij} + S_{ij}}{(\lambda_2 - \lambda_1)(\lambda_3 - \lambda_1)} \tag{56}$$

$$(Z_{ij})_2 = \frac{\delta_{ij}\lambda_1\lambda_3 - (\lambda_1 + \lambda_3)D_{ij} + S_{ij}}{(\lambda_1 - \lambda_2)(\lambda_3 - \lambda_2)} \tag{57}$$

$$(Z_{ij})_3 = \frac{\delta_{ij}\lambda_1\lambda_2 - (\lambda_1 + \lambda_2)D_{ij} + S_{ij}}{(\lambda_1 - \lambda_3)(\lambda_2 - \lambda_3)} \tag{58}$$

where

$$S_{ij} = \sum_l D_{il}D_{lj} \tag{59}$$

When these expressions are substituted into Eq. 5, formulas for $c_i - \bar{c}_i$ are obtained. These $(Z_{ij})_k$ expressions can be substituted into Eq. 18, but the results are not in Kim's forms(5). However they can be transformed into Kim's quantities. Consequently, Kim's solutions for the quaternary D_{ij} (5) can be used. The transformations involve the traces of $\mathbf{D}$(13); namely,

$$T_1 = \lambda_1 + \lambda_2 + \lambda_3 = D_{11} + D_{12} + D_{13} \tag{60}$$

$$T_2 = \lambda_1\lambda_2 + \lambda_2\lambda_3 + \lambda_3\lambda_1 = D_{11}D_{22} - D_{12}D_{21} + D_{22}D_{33} - D_{23}D_{32} + D_{33}D_{11} - D_{31}D_{13} \tag{61}$$

$$T_3 = \lambda_1\lambda_2\lambda_3 = \det(\mathbf{D}) \tag{62}$$

The expression for Γ_k becomes

$$\Gamma_k = \frac{\Sigma_Y + \lambda_k \Sigma_X - \lambda_k(\lambda_m + \lambda_n)}{(\lambda_m - \lambda_k)(\lambda_n - \lambda_k)} \qquad m \neq n \neq k \tag{63}$$

where Σ_X and Σ_Y are

$$\Sigma_X = \sum_j X_j \alpha_j \tag{64}$$

$$\Sigma_Y = \sum_j Y_j \alpha_j \tag{65}$$

and X_j is given by

$$R_j X_j = \sum_i R_i D_{ij} \tag{66}$$

Here j is a running index, not a fringe number.

The expressions for the Y_j in terms of the D_{ij} are less easy to systematize, but are given by Kim(5). Our expressions for X_j and Y_j are actually T_3 times the corresponding X_j and Y_j expressions of Kim(5).

With Eq. 63, the Rayleigh expression for the reduced fringe number f(j) becomes suitable for non-linear least-squaring:

$$f(j) = \frac{[\Sigma_Y + \lambda_1 \Sigma_X - \lambda_1(\lambda_2 + \lambda_3)]\,\mathrm{erf}(y_j/\lambda_1^{1/2})}{(\lambda_2 - \lambda_1)(\lambda_3 - \lambda_1)}$$

$$+ \frac{[\Sigma_Y + \lambda_2 \Sigma_X - \lambda_2(\lambda_1 + \lambda_3)]\,\mathrm{erf}(y_j/\lambda_2^{1/2})}{(\lambda_1 - \lambda_2)(\lambda_3 - \lambda_2)} \tag{67}$$

$$+ \frac{[\Sigma_Y + \lambda_3 \Sigma_X \quad \lambda_3(\lambda_1 + \lambda_2)]\,\mathrm{erf}(y_j/\lambda_3^{1/2})}{(\lambda_1 - \lambda_3)(\lambda_2 - \lambda_3)}$$

Here j refers to the fringe number.

Since the fringe number j, the total number of fringes J, and the x_j and t of y_j are the experimental quantities, Σ_Y, Σ_X, and the λ_k can be obtained from a single experiment. Three experiments with different α_i ratios are required to isolate the individual X_i and Y_i. Consequently, when data from the three (or more) such experiments are least-squared together, the 9 least-square parameters X_i, Y_i, and λ_i are sufficient to determine the 9 D_{ij}.

The solution for the D_{ij} in terms of our X_i, Y_i, and the λ_i can be obtained from Kim's solution(5). The result has a number of equivalent forms, but in terms of the traces T_i (Eqs. 60-62) is

$$\frac{R_i D_{ij}}{R_j} = \frac{(Y_k - Y_l)X_j T_1 - (Y_k - Y_l)T_2 - (X_k - X_l)T_3 + (Y_k - Y_l)Y_j + (X_k Y_l - X_l Y_k)X_j}{X_1(Y_2 - Y_3) + X_2(Y_3 - Y_1) + X_3(Y_1 - Y_2)} \tag{68}$$

where there is a cyclic permutation of the indices; namely, if i = 1, then k = 2, l = 3; if i = 2, then k = 3, l = 1; and if i = 3 then k = 1, l = 2.

In reference 7, the X_i and Y_i were replaced in f(j) by a_i and b_i whose expressions are the appropriate coefficients of α_i in Eq. 63 minus the appropriate term $\lambda_k(\lambda_m + \lambda_n)\alpha_i$, and the solution for D_{ij} was given in terms of a_i, b_i, and λ_i. The a_i, b_i expressions are more convenient for the distinct eigenvalue case, but are not suited for the use of L'Hôpital's rule to examine the equal eigenvalue cases.

We are now in a position to obtain the results for the two cases of equal eigenvalues. These can be obtained from L'Hôpital's rule on the f_{ij}, after bringing the expression for the distinct case to a common denominator. Similarly, L'Hôpital's rule can be applied to the expression for f(j) in the distinct case, again after bringing f(j) to a common denominator.

B. Two Eigenvalues Equal

To be definite, we take $\lambda_2 = \lambda_3$. The application of L'Hôpital's rule to f_{ij} is by differentiating numerator and denominator with respect to λ_3 and then setting $\lambda_3 = \lambda_2$. The results are as follows.

$$F_1 = \mathrm{erf}(y_j/\lambda_1^{1/2}) \tag{69}$$

$$F_2 = \mathrm{erf}(y_j/\lambda_2^{1/2}) \tag{70}$$

$$F_3 = F_2' = \frac{\partial F_2}{\partial \lambda_2} = \frac{-y_j \exp(-y_j^2/\lambda_2^{1/2})}{\pi^{1/2}\lambda_2^{3/2}} \tag{71}$$

The expressions for $(Z_{ij})_k$ are

$$(Z_{ij})_1 = \frac{\delta_{ij}\lambda_2^2 - 2\lambda_2 D_{ij} + S_{ij}}{(\lambda_2 - \lambda_1)^2} \tag{72}$$

$$(Z_{ij})_2 = \frac{\delta_{ij}(\lambda_1^2 - 2\lambda_1\lambda_2) + 2\lambda_2 D_{ij} - S_{ij}}{(\lambda_2 - \lambda_1)^2} \tag{73}$$

$$(Z_{ij})_3 = \frac{\delta_{ij}\lambda_1\lambda_2 - (\lambda_1 + \lambda_2)D_{ij} + S_{ij}}{(\lambda_2 - \lambda_1)} \tag{74}$$

In terms of the X_i and Y_i,

$$f(j) = \frac{[\Sigma_Y + \lambda_1\Sigma_X - 2\lambda_1\lambda_2]}{(\lambda_2 - \lambda_1)^2} . F_1$$

$$- \frac{[\Sigma_Y + \lambda_1\Sigma_X - (\lambda_1^2 + \lambda_2^2)]}{(\lambda_2 - \lambda_1)^2} . F_2 \tag{75}$$

$$+ \frac{[\Sigma_Y - \lambda_2\Sigma_X - \lambda_2(\lambda_1 + \lambda_2)]}{(\lambda_2 - \lambda_1)} . F_2'$$

The least-square variables in this case are exactly the same as in the distinct case. Only the coefficients and one function are different. Therefore the D_{ij} solution (Eq. 68) is the same except we set $\lambda_3 = \lambda_2$ in the traces T_i. Again at least three different experiments with different α_i ratios are required.

C. All Three Eigenvalues Are Equal

We can start L'Hôpital's rule with the two equal eigenvalues case. We first bring f_{ij} (or f(j)) to a common denominator. It is necessary to differentiate twice because of the $(\lambda_2 - \lambda_1)^2$ in the denominator. Differentiation with either λ_1 or λ_2 is satisfactory, but the steps are easier with λ_1. Then λ_2 is set equal to λ_1. The results are as follows

$$F_1 = \mathrm{erf}(y_j/\lambda_1^{1/2}) \tag{76}$$

$$F_2 = F_1' = \frac{\partial F_1}{\partial \lambda_1} = \frac{-y_j \exp(-y_j^2/\lambda_1^{1/2})}{\pi^{1/2}\lambda_1^{3/2}} \tag{77}$$

$$F_3 = F_1'' = \frac{\partial^2 F_1}{\partial \lambda_1^2} = \frac{-y_j(2y_j^2 - 3\lambda_1)\exp(-y_j^2/\lambda_1)}{2\pi^{1/2}\lambda_1^{7/2}} \tag{78}$$

The expressions for $(Z_{ij})_k$ are

$$(Z_{ij})_1 = \delta_{ij} \tag{79}$$

$$(Z_{ij})_2 = -(\delta_{ij}\lambda_1 - D_{ij}) \tag{80}$$

$$(Z_{ij})_3 = \frac{\delta_{ij}\lambda_1^2 - 2\lambda_1 D_{ij} + S_{ij}}{2} \tag{81}$$

The expression for f(j) in terms of X_i and Y_i is

$$f(j) = F_1 + (\Sigma_X - \lambda_1)F_1' + \frac{(\Sigma_Y + \lambda_1\Sigma_X - 2\lambda_1^2)F_1''}{2} \tag{82}$$

As in the distinct and two equal eigenvalue cases, the least-square variables X_i and Y_i are the same but the functions of λ_i (in this case λ_1 only) are different. At least three different experiments with three different α_i ratios are still required. The D_{ij} solution (Eq. 68) is also the same, except that $\lambda_3 = \lambda_2 = \lambda_1$ are substituted in the expressions for the traces T_i.

Discussion

As noted earlier, explicit solutions of the diffusion equation have been available for ternary(1,2) and quaternary systems(4,5) for distinct eigenvalues. Expressions relating least-square parameters to experimental quantities and expressions relating the diffusion coefficients to the least-square parameters are available for ternary and quaternary Gouy and Rayleigh systems(7). In addition, the solution of the diffusion equation for the ternary equal eigenvalue case was given by Toor(4).

Our new results include the solutions of the diffusion equation for the quaternary system in terms of the $(Z_{ij})_k$ for both of the equal eigenvalue cases (2 equal and all 3 equal). In addition, we have given the expressions connecting the least-square parameters to experimental quantities and connecting the diffusion coefficients to the least-square parameters for ternary Rayleigh and Gouy systems and quaternary Rayleigh systems.

In all these cases, whether the eigenvalues are distinct or equal, the experimental quantities and least-square parameters are the same for a given number of components. Only their combinations and the functions of λ_i are different. The expressions for the D_{ij} in terms of the least-square parameters are also the same for a given number of components, except that the traces T_i are evaluated with either all λ_i distinct or with one or more of $\lambda_m = \lambda_n$. These same conclusions apply to other experimental situations, such as restricted diffusion or the diaphragm cell.

The results sketched here are thus complete for Gouy and Rayleigh interferometry for ternary systems and for Rayleigh for quaternary systems. Full details, alternate expressions, and Gouy results for quaternary systems will be presented elsewhere.

References

1. H. Fujita and L. J. Gosting, J. Am. Chem. Soc. 78, 1099 (1956).
2. J. S. Kirkaldy, Can. J. Phys. 36, 899 (1958).
3. J. S. Kirkaldy, Can J. Phys. 37, 30 (1959).
4. H. Toor, AIChE J. 10, 460 (1964).
5. H. Kim, J. Phys. Chem. 70, 562 (1966).
6. H. Fujita and L. J. Gosting, J. Phys. Chem. 64, 1256 (1960).
7. D. G. Miller, J. Phys. Chem. 92, 4222 (1988).
8. L.-O. Sundelöf and I. Södervi, Ark. Kemi 21, 143 (1963).
9. D. G. Miller, V. Vitagliano, and R. Sartorio, J. Phys. Chem. 90, 1509 (1986).
10. H. J. V. Tyrrell and K. R. Harris, Diffusion in Liquids, Butterworths, London (1984).
11. L. J. Gosting and L. Onsager, J. Am. Chem. Soc. 74, 6066 (1952).
12. J. G. Albright and D. G. Miller, J. Phys. Chem. 93, 2169 (1989).
13. J. S. Kirkaldy, D. Weichert, and Zia-Ul-Haq, Can. J. Phys. 41, 2166 (1963).
14. D. G. Miller, J. Solution Chem. 10, 831 (1981).

Initial-value solutions for spinodal decomposition in ternary solids

J.S. Kirkaldy, K. Balasubramanian
Institute for Materials Research, McMaster University, Hamilton, Ontario, Canada, L8S 4M1

Y.J.M. Brechet
L.T.P.C.M./E.N.S.E.E.G., Domaine Universitaire, St-Martin d'Hères, Cedex, France

Abstract

The establishment of a Fourier decomposable initial value representation of spinodal decomposition in a ternary alloy requires that the 2×2 diffusion (D) and gradient energy (G) matrices be simultaneously brought to the diagonal, or equivalently that they commute. The three sub-matrices involving solution thermodynamics, mobility and gradient terms are examined within standard atomic models with the view to establishing conditions for commutability within shallow spinodal states. Establishing that this condition is thermodynamically and kinetically feasible, though far from general, we explore the special case where the G-matrix is homothetic, implying that G is diagonal with equal eigenvalues. Within this specialization we demonstrate the trend whereby composition vector components of arbitrary fluctuations which parallel the near critical point tielines rapidly dominate the growth manifold, and this in parallel with the expected independent trend for fastest growing wave number selection.

Introduction

The experimental record for spinodal decomposition pertains as often to ternary as it does to binary solids so it appears worthwhile to expand on our earlier initial value closed form solutions of the reaction equations (Kirkaldy and Purdy; 1), by including strain and gradient energy terms. Our approach contrasts with that of de Fontaine (2) and Morral and Cahn (3) where the matrix eigen-values are functions of the wave number β thus precluding the expression of a closed form initial value problem. The linearized theory consists of a fourth order time-dependent diffusion equation (de Fontaine; 2) which contains a 2×2 matrix of modified diffusion coefficients (second spatial order) and a 2×2 matrix of modified gradient energy coefficients (fourth spatial order), viz.,

$$\frac{\partial C_A}{\partial t} = D_{AA}\nabla^2 C_A + D_{AB}\nabla^2 C_B - G_{AA}\nabla^4 C_A - G_{AB}\nabla^4 C_B \tag{1}$$

$$\frac{\partial C_B}{\partial t} = D_{BA}\nabla^2 C_A + D_{BB}\nabla^2 C_B - G_{BA}\nabla^4 C_A - G_{BB}\nabla^4 C_B \tag{2}$$

As it will become apparent the specification of an analytic or closed form Fourier decomposable initial value problem requires that the matrices D and G be simultaneously brought to the diagonal. That is to say if equations (1) and (2) in one dimension can be transformed to

$$\frac{\partial \bar{C}_A}{\partial t} = D_A \frac{\partial^2 \bar{C}_A}{\partial x^2} - G_A \frac{\partial^4 \bar{C}_A}{\partial x^2} \tag{3}$$

and

$$\frac{\partial \bar{C}_B}{\partial t} = D_B \frac{\partial^2 \bar{C}_B}{\partial x^2} - G_B \frac{\partial^4 \bar{C}_B}{\partial x^2} \tag{4}$$

by the linear transformations defined by the 2×2 matrix α^{-1}, viz.,

$$\overline{C}_A = \alpha^{-1}_{AA} C_A + \alpha^{-1}_{AB} C_B \tag{5}$$

and

$$\overline{C}_B = \alpha^{-1}_{BA} C_A + \alpha^{-1}_{BB} C_B \tag{6}$$

Then we can express the solution vector (C_A, C_B) in the Fourier integral form

$$\begin{pmatrix} C_A - C_{AO} \\ C_B - C_{BO} \end{pmatrix} = \alpha \begin{bmatrix} \int_{-\infty}^{\infty} \overline{C}_A(\beta) \cos\beta x \exp[-(\beta^2 D_A + \beta^4 G_A)t] \, d\beta \\ \\ \int_{-\infty}^{\infty} \overline{C}_B(\beta) \cos\beta x \exp[-(\beta^2 D_B + \beta^4 G_B)t] \, d\beta \end{bmatrix} \tag{7}$$

and evaluate the eigenvector (C_A, C_B) by taking the Fourier transform of ($C_A(t=0) - C_{AO}$, $C_B(t=0) - C_{BO}$). In view of the known properties of D and G this does not seem to be in general possible. In the first section to follow we state the matrix restrictions on the problem of finding simple initial value solutions, in the second section we examine the D and G matrices from the point-of-view of a regular solution thermodynamic model and a common kinetic approximation which subsumes spinodal decomposition and their potential for diagonalization, and the third section deals with the early stage evolution of a model system on the Gibbs Triangle as an initial value problem..

The Matrix Restrictions

As we shall now demonstrate, for the two real matrices D and G to be precisely brought to the diagonal it is necessary and sufficient that they commute, viz.,

$$DG = GD \tag{8}$$

Consider that if D and G are simultaneously diagonalizable then there is a unitary matrix S ($SS^{-1} = I$) such that

$$D' = S^{-1}DS \quad \text{and} \quad G' = S^{-1}GS \tag{9}$$

where D′ and G′ commute since they are diagonal. Hence

$$D'G' = S^{-1}DS \cdot S^{-1}GS = S^{-1}DGS \tag{10}$$

$$G'D' = S^{-1}GS \cdot S^{-1}DS = S^{-1}GDS \tag{11}$$

which argument can be reversed and the proof is complete.

Now it is known that D and G are both products of symmetric matrices (see next section) of the form

$$\rho D = L\mu \quad \text{and} \quad \rho G = 2L\chi \tag{12}$$

where ρ is the density of the ternary solution. We now ask under what conditions on the constituent matrices D and G commute. First define

$$\mu A = \chi \tag{13}$$

and note the consequences of commutation

$$L\chi L\mu = L\mu L\chi \tag{14}$$

or

$$\chi L\mu = \mu L\chi \tag{15}$$

or

$$\chi L\mu = \mu L A\mu \tag{16}$$

Thus the symmetry relation $\chi = {}^t\chi$ implies

$$^{t}(A\mu) = {}^{t}\mu\,{}^{t}A = A\mu = \chi = \mu\,{}^{t}A \tag{17}$$

whence

$$\mu LA\mu = \mu\,{}^{t}AL\mu \tag{18}$$

or

$${}^{t}AL = LA \tag{19}$$

But

$$^{t}(LA) = {}^{t}A\,{}^{t}L = {}^{t}AL \tag{20}$$

Therefore $L\chi$ and $L\mu$ commute if and only if LA is symmetric, viz.,

$$(L\chi\mu^{-1}) = {}^{t}(L\chi\mu^{-1}) \tag{21}$$

Since the argument is symmetric between μ and χ we also have

$$(L\mu\chi^{-1}) = {}^{t}(L\mu\chi^{-1}) \tag{22}$$

The corollary sufficient conditions for commutation are thus:

(1) Either D or G are diagonal

(2) $\chi\mu^{-1}$ or $\mu\chi^{-1}$ is diagonal which imply χ proportional to μ

(3) $L\mu^{-1}$, $L\chi$ and $\chi\mu^{-1}$ are simultaneously symmetric

As we shall see the physics usually denies conditions (2) and (3) and the diagonality of D so we must examine the implications of diagonal G.

If we explicitly construct the commutation relation (8) we obtain the conditions in terms of the matrix elements

$$\frac{D_{BB} - D_{AA}}{G_{BB} - G_{AA}} = \frac{D_{AB}}{G_{AB}} = \frac{D_{BA}}{G_{BA}} \tag{23}$$

Thus is $G_{AB} = G_{BA} = 0$ then $G_{BB} = G_{AA}$ and the eigenvalues

$$G_A = G_B = G_{AA} = G_{BB} = G \tag{24}$$

Matrix G is said to be homothetic, which is a very strong condition indeed. We shall now investigate the component matrices L, μ and χ more fully.

Restrictions for a Realizable Model

For a typical z-coordinated nearest neighbour cubic solid the Bragg-Williams model for a ternary A, B, C solution of mole number $n = n_A + n_B + n_C$ the Helmholtz Free Energy $A = E - TS$ in terms of pair energies ε is (Brown and Kirkaldy; 4)

$$A = A^{o} + \frac{zn_A^2}{2n}[\varepsilon_{AA} + \varepsilon_{CC} - 2\varepsilon_{AC}] + \frac{zn_B^2}{2n}[\varepsilon_{AA} + \varepsilon_{CC} - 2\varepsilon_{AC}]$$

$$+ z\frac{n_A n_B}{n}[\varepsilon_{AB} + \varepsilon_{CC} - \varepsilon_{BC} - \varepsilon_{AC}] + zn_A[\varepsilon_{AC} - \varepsilon_{CC}] + zn_B[\varepsilon_{BC} - \varepsilon_{CC}]$$

$$- T[n_A S_A^o + n_B S_B^o + n_C S_C^o - kn_A \ln(n_A) - kn_B \ln(n_B) - kn_C \ln(n_C)] \tag{25}$$

By definition the chemical potentials are

$$\mu_i = N_A\,\partial A/\partial n_i \tag{26}$$

where N_A is Avogadro's number, whence choosing A and B for brevity and clarity as the most dilute species ($X_A + X_B < 0.3$) we obtain the chemical potentials in terms of mole fractions

$$X_i = n_i/(n_A + n_B + n_C) \tag{27}$$

$$\mu_A = \mu_A^o + RT\ln(X_A/X_C) + zN_A X_A[\varepsilon_{AA} + \varepsilon_{CC} - 2\varepsilon_{AC}] + zN_A X_B[\varepsilon_{AB} + \varepsilon_{CC} - \varepsilon_{AC} - \varepsilon_{BC}] \tag{28}$$

$$\mu_B = \mu_B^o + RT\ln(X_B/X_C) + zN_A X_A[\varepsilon_{AB} + \varepsilon_{CC} - \varepsilon_{AC} - \varepsilon_{BC}] + zN_A X_B[\varepsilon_{BB} + \varepsilon_{CC} - 2\varepsilon_{BC}] \tag{29}$$

where the symmetry is a thermodynamic imperative (Maxwell's equation) and for the solvent

$$\mu_C \simeq \mu_C^o + RT\ln X_C \tag{30}$$

The D matrix is very closely approximated by Eq. (12) where L is the positive definite Onsager mobility matrix and μ, which is a specialization of the Hessian of the Helmholtz Free Energy, incorporates

$$\mu'_{ik} = \partial(\mu_i - \mu_c)/\partial X_k \tag{31}$$

and is therefore specified from equations (28) - (30),

$$\mu'_{AA} \simeq \frac{RT}{X_A} - 2zN_A\left[\varepsilon_{AC} - \frac{\varepsilon_{AA} + \varepsilon_{CC}}{2}\right] \tag{32}$$

$$\mu'_{BB} \simeq \frac{RT}{X_B} - 2zN_A\left[\varepsilon_{BC} - \frac{\varepsilon_{BB} + \varepsilon_{CC}}{2}\right] \tag{33}$$

$$\mu'_{AB} = \mu'_{BA} = -zN_A[\varepsilon_{AC} + \varepsilon_{BC} - \varepsilon_{AB} - \varepsilon_{CC}] \tag{34}$$

There are also additive strain energy terms which will be discussed briefly at a later stage (see Eqs. 40-43)

The χ matrix is directly related to the gradient energy and can be evaluated for the regular solution model by a method proposed by Becker (5). Consider Fig. 1 and let X'_A, X'_B, X'_C and X''_A, X''_B, X''_C be the near compositions of the two representative matrix phases of lattice parameter a.

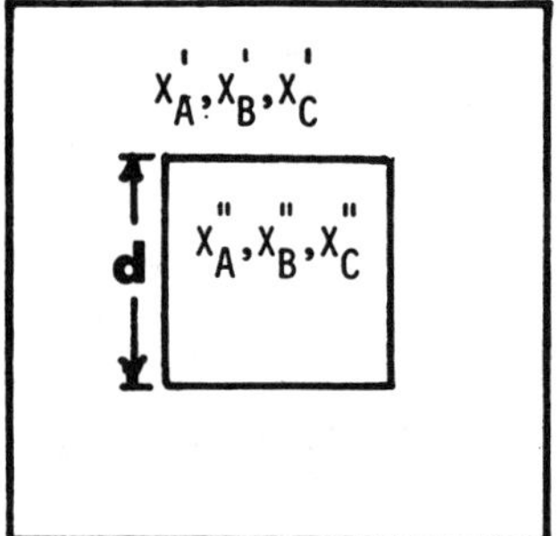

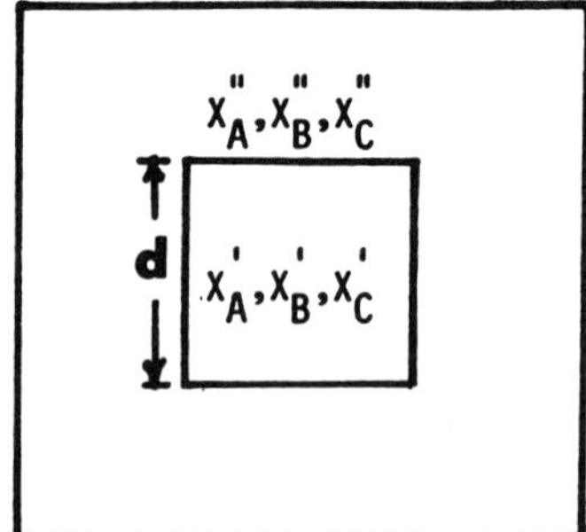

Fig. 1 Construction for evaluating the gradient energy matrix in a ternary system.

Interchanging a small cube of dimensions d of the above compositions results in the increase in the surface free energy of the system which is twice that due to presence of the initial inhomogeneity, making a zero change in the total volume free energy. Here (ξ/a^2) = total number of bonds crossing the interface per m^2; d = width of the small cubic nucleus; $N = 6d^2\,(\xi/a^2)$ = total number of bonds crossing the interface for each cube and $12d^2$ is the total interface area created (two cubes taken together). Evaluation of the fraction of bonds of each type crossing the interface of each cube yields

	A−A	B−B	C−C	A−B	B−C	C−A
Left side before	$X'_A X'_A$	$X'_B X'_B$	$X'_C X'_C$	$2X'_A X'_B$	$2X'_B X'_C$	$2X'_C X'_A$
Left side after	$X'_A X''_A$	$X'_B X''_B$	$X'_C X''_C$	$(X'_A X''_B + X''_A X'_B)$	$(X'_B X''_C + X''_B X'_C)$	$(X'_C X''_A + X''_C X'_A)$
Right side before	$X''_A X''_A$	$X''_B X''_B$	$X''_C X''_C$	$2X''_A X''_B$	$2X''_B X''_C$	$2X''_C X''_A$
Right side after	$X''_A X'_A$	$X''_B X'_B$	$X''_C X'_C$	$(X''_A X'_B + X'_A X''_B)$	$(X'_B X''_C + X''_B X'_C)$	$(X'_C X''_A + X''_C X'_A)$

The change in the number of bonds of each kind due to interchange is given by

A−A $\quad (X'_A X''_A + X''_A X'_A - X'_A X'_A - X''_A X''_A)N = -(X''_A - X'_A)^2 \cdot N$

B−B $\quad (X'_B X''_B + X''_B X'_B - X'_B X'_B - X''_B X''_B)N = -(X''_B - X'_B)^2 \cdot N$

C−C $\quad (X'_C X''_C + X''_C X'_C - X'_C X'_C - X''_C X''_C)N = -(X''_C - X'_C)^2 \cdot N$

A−B $\quad [2(X'_A X''_A + X'_B X''_A) - 2X'_A X'_B - 2X''_A X''_B]N = 2(X'_A - X''_A)(X''_B - X'_B) \cdot N$

B−C $\quad [2(X'_B X''_C + X'_C X''_B) - 2X'_B X'_C - 2X''_B X''_C]N = 2(X'_B - X''_B)(X''_C - X'_C) \cdot N$

C−A $\quad [2(X'_C X''_A + X'_A X''_C) - 2X'_C X'_A - 2X''_C X''_A]N = 2(X'_C - X''_C)(X''_A - X'_A) \cdot N$

Eliminating X_C via $X_C = 1 - X_A - X_B$ in the above equations and assigning ε_{AA}, ε_{BB}, ε_{CC} etc., as the pair interaction energies as before, we obtain for the free energy change per unit volume (neglecting entropy change)

$$\Delta A = \xi N_A \rho \left[\left[\varepsilon_{CA} - \frac{\varepsilon_{AA} + \varepsilon_{CC}}{2} \right] (X''_A - X'_A)^2 + \left[\varepsilon_{BC} - \frac{\varepsilon_{BB} + \varepsilon_{CC}}{2} \right] (X''_B - X'_B)^2 + 2 \left[\frac{\varepsilon_{CA} + \varepsilon_{BC} - \varepsilon_{AB} - \varepsilon_{CC}}{2} \right] \cdot (X''_A - X'_A)(X''_B - X'_B) \right] \tag{35}$$

Dividing and multiplying by a^2 to introduce gradients one can recognize that as usually defined χ is the coefficient matrix in Eq. 35 multipl:ed by a^2.

To obtain D and G both μ (incorporating strain energy) and χ must be multiplied by the Onsager mobility matrix (Eqs. 7). Neglecting correlation effects, which amount to a maximum of about 10% in the D values for face-centred cubic materials and imposing a vacancy diffusion model, the L-matrix is (Kirkaldy and Young; 6)

$$L_{AA} = \frac{\rho a^2}{RT}[P_AX_A(1-2X_A) + X_A^2(P_AX_A + P_BX_B + P_CX_C)] \tag{36}$$

$$L_{BB} = \frac{\rho a^2}{RT}[P_BX_B(1-2X_B) + X_B^2(P_AX_A + P_BX_B + P_CX_C)] \tag{37}$$

$$L_{AB} = L_{BA} = \frac{\rho a^2}{RT}[P_AX_A^2X_B + P_BX_AX_B^2 - X_AX_B(P_AX_A + P_BX_B + P_CX_C)] \tag{38}$$

where the jump rates or probabilities/unit time are

$$P_i \simeq D_i^*/\rho a^2 \tag{39}$$

the D_i^* being the tracer coefficients. The latter may be regarded in the first instance as being independent of the thermodynamic parameters (see, however, Brown and Kirkaldy; 4).

Now given the μ, χ and L matrices we can write out the 2×2 D and G matrices in full (de Fontaine; 2).

$$\rho D_{AA} = L_{AA}(\mu'_{AA} + \phi\eta_A^2) + L_{AB}(\mu'_{BB} + \phi\eta_A\eta_B) \tag{40}$$

$$\rho D_{AB} = L_{AA}(\mu'_{AB} + \phi\eta_A\eta_B) + L_{AB}(\mu'_{AB} + \phi\eta_B^2) \tag{41}$$

$$\rho D_{BB} = L_{BA}(\mu'_{AA} + \phi\eta_A^2) + L_{BB}(\mu'_{BA} + \phi\eta_A\eta_B) \tag{42}$$

$$\rho D_{BA} = L_{BA}(\mu'_{AB} + \eta_A\eta_B) + L_{BB}(\mu'_{BB} + \phi\eta_B^2) \tag{43}$$

where η_A and η_B are isotropic Vegard's Law strains, and

$$\rho G_{AA} = 2(L_{AA}\chi_{AA} + L_{AB}\chi_{BA}) \quad ; \quad \rho G_{BA} = 2(L_{BA}\chi_{AA} + L_{BB}\chi_{BA}) \tag{44}$$

$$\rho G_{AB} = 2(L_{AA}\chi_{AB} + L_{AB}\chi_{BB}) \quad ; \quad \rho G_{BB} = 2(L_{BA}\chi_{AB} + L_{BB}\chi_{BB}) \tag{45}$$

Note that the elastic strain energy matrix is added directly to the chemical matrix μ'_{ik} in a way which sustains the symmetry. The onset of spinodal decomposition is defined by the vanishing of the determinant of the μ matrix, viz.,

$$\mathrm{Det}\mu = \mu'_{AA}\mu'_{BB} - \mu'^2_{AB} + \phi^2(\eta_A^2\mu'_{BB} - 2\mu'_{AB}\eta_A\eta_B + \eta_B^2\mu'_{AA}) \tag{46}$$

When Det μ' vanishes the right hand side is a perfect square implying that an enhanced driving force which is independent of the wave number is required for onset of the reaction. This is equivalent to a uniform depression of the miscibility gap, so we lose nothing in principle by dropping the strain energy term from further consideration ($\phi = 0$; $\mu = \mu'$).

It now becomes clear for this dilute regular solution approximation, that although closely related, μ, χ are not proportional on account of the leading concentration dependent terms in equations (32) and (33) which must be of the same order of magnitude as the non-dependent terms. To see this clearly examine the binary limit $X_B \rightarrow 0$ whence for the critical point $\mu'_{AA} = 0$ yields the binary relation at $X_A = X_C = 0.5$,

$$RT_C = \frac{N_Az}{2}\left(\varepsilon_{AC} - \frac{\varepsilon_{AA} + \varepsilon_{CC}}{2}\right) \tag{47}$$

Furthermore, on account of the multiple parameter dimensionality within D and G mutual symmetry would be fortuitous. Finally, it would be equally fortuitous that $\rho D = L\mu$ is diagonal in view of the strong concentration dependencies of both of the factor matrices. It remains therefore to investigate whether diagonality of $\rho G = L\chi$ is a feasible or probable eventuality, which is to say

$$L_{AA}\chi_{AB} + L_{AB}\chi_{BB} = 0 \quad \text{and} \quad L_{BA}\chi_{AA} + L_{BB}\chi_{BA} = 0 \tag{48}$$

implying

$$L_{AA}\chi_{AA} + L_{AB}\chi_{BA} = L_{BA}\chi_{AB} + L_{BB}\chi_{BB} \tag{49}$$

Here we refer to a common ternary miscibility gap on the Gibbs Triangle as indicated schematically in Fig. 2. We assume the absence of ordering between the pairs AC and BC which can be taken to imply that AC and BC are moderately repulsive relative to AA and BB and AB pairs are more strongly repulsive (ε is positive for unlike pairs). Referring to the χ elements of Eq. 35 we can define the standard state of C such that $\varepsilon_{CC} = 0$.

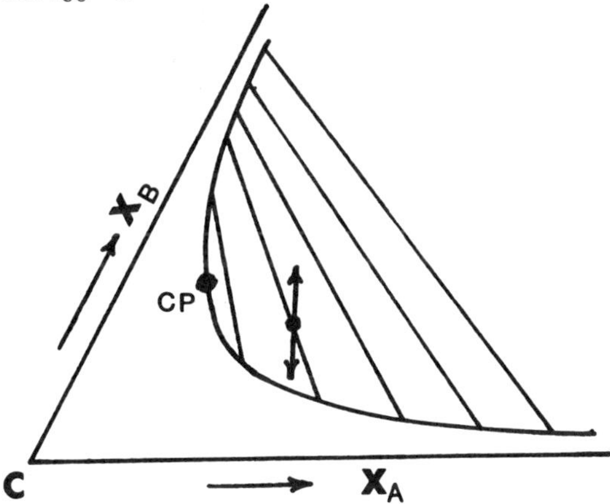

Fig. 2 Schematic miscibility gap on the Gibbs Triangle identifying a critical point and a test fluctuation.

As indicated earlier we consider a relatively dilute spinodal alloy X_A, X_B (Fig. 2) so we can approximate the L matrix from Eqs. 36 to 38 as

$$L_{AA} \simeq \frac{\rho a^2}{RT} P_A X_A \; ; \; L_{BB} \simeq \frac{\rho a^2}{RT} P_B X_B \; ; \; L_{AB} = L_{BA} \simeq -\frac{\rho a^2}{RT} P_C X_A X_B \tag{50}$$

Then from Eqs. 48

$$P_A[\frac{1}{2}(\varepsilon_{AC} + \varepsilon_{BC} - \varepsilon_{AB})] = P_C X_B(\varepsilon_{BC} - \frac{1}{2}\varepsilon_{BB}) \tag{51}$$

and

$$P_B[\frac{1}{2}(\varepsilon_{AC} + \varepsilon_{BC} - \varepsilon_{AB})] = P_C X_A(\varepsilon_{CA} - \frac{1}{2}\varepsilon_{AA}) \tag{52}$$

and from Eq. 49, which follows from Eqs. 48,

$$G_A = 2\xi X_A \frac{\rho^2 a^2}{RT}\left[P_A(\varepsilon_{AC} - \frac{\varepsilon_{AA}}{2}) - P_C X_B \cdot \frac{1}{2}(\varepsilon_{AC} + \varepsilon_{BC} - \varepsilon_{AB})\right]$$

$$\simeq 2\xi X_B \frac{\rho^2 a^2}{RT}[P_B(\varepsilon_{BC} - \frac{\varepsilon_{BB}}{2}) - P_C X_A \cdot \frac{1}{2}(\varepsilon_{AC} + \varepsilon_{BC} - \varepsilon_{AB})] = G_B \quad (53)$$

For the sake of argument consider first a kinetically near symmetric solution ($P_A \sim P_B \sim P_C$) with $\varepsilon_{AC} + \varepsilon_{BC}$ only slightly greater than ε_{AB}. We are always free to choose X_A, X_B so that Eqs. 51, 52 and 53 are satisfied. Furthermore, in this case the trailing terms in G_A and G_B vanish relative to the leading terms so again choosing X_A, X_B appropriately we can assume

$$G_A \simeq 2\xi \frac{\rho^2 a^2}{RT} P_A X_A (\varepsilon_{AC} - \frac{\varepsilon_{AA}}{2}) \simeq 2\xi \frac{\rho^2 a^2}{RT} P_B X_B (\varepsilon_{BC} - \frac{\varepsilon_{BB}}{2}) \simeq G_B \quad (54)$$

which remarkably does not depend sensitively on gradient terms associated with the strongest part of the spinodal segregation designated by $\varepsilon_{AB} > 0$. That is to say, as regards to gradient energy the system acts like independent binaries (AC and BC). The reason lies in the fact that for the symmetric kinetics considered in a relatively dilute solution AB exchange by the intermediary of vacancies is a rare event. It remains to check whether a relation of form (54) can obtain near a critical point configuration like Fig. 2.

Proceeding from the vanishing of the quadratic μ_{ik} determinant to the same approximations we obtain for the critical temperature (cf. Eq. 47)

$$RT_C = 2zN_A X_A(\varepsilon_{AC} - \varepsilon_{AA}/2) + zN_A X_A X_B(\varepsilon_{AC} + \varepsilon_{BC} - \varepsilon_{AB})^2/2[X_A(\varepsilon_{AC} - \varepsilon_{AA}/2) - X_B(\varepsilon_{BC} - \varepsilon_{BB}/2)] \quad (55)$$

Now in Kirkaldy and Purdy (1) we established the exact expression for the slope of the limiting tie-line at the critical point

$$\delta = (dX_B/dX_A)_C = -\mu_{AA}/\mu_{AB} \quad (56)$$

Thus from relations (32), (34) and (56) we obtain

$$(dX_B/dX_A)_C = X_B(\varepsilon_{AC} + \varepsilon_{BC} - \varepsilon_{AB})/2[X_A(\varepsilon_{AC} - \varepsilon_{AA}/2) - X_B(\varepsilon_{BC} - \varepsilon_{BB}/2)] \quad (57)$$

the denominator of which should be negative to accord with Fig. 2. Referring to Eq. 54 we see that G can be homothetic for systems with a wide range of relative mobilities and (or) interaction parameters. We could, of course, substitute numerical integration where all the specializations, restrictions and approximations can be dropped, but hereby we would lose transparency and pedagogic value.

The foregoing illustrates what is insufficiently realized, even for the binary case; that the thermodynamics and kinetics of reaction near a critical point are strongly coupled and one ignores this by arbitrary choice of parameters at peril of gross error.

Initial Value Solutions and Their Evolution on the Gibbs Triangle

As stated earlier, de Fontaine, and Morral and Cahn have examined a class of formal solutions of Eqs. (1) and (2) which appear to avoid the diagonalization problem. However, Morral and Cahn's suggested procedure of Fourier analysis of the initial conditions is at best achievable through a large number of iterative operations on transcendentals so is in effect a modified procedure of numerical analysis. Our method with limited but sufficient generality proceeds to a closed solution via conventional Fourier integral methods.

Proceeding by substitution of Eqs. 7 in (1) and (2) one obtains the characteristic equations

$$\frac{(D_B - D_{AA})}{D_{AB}} = \frac{D_{AB}}{(D_B - D_{BB})} = \frac{\alpha_{BB}}{\alpha_{AB}} = \frac{(G_A - G_{AA})}{G_{AB}} = \frac{G_{BA}}{(G_B - G_{BB})} \quad (56)$$

and

$$\frac{(D_A - D_{BB})}{D_{BA}} = \frac{D_{AB}}{(D_A - D_{AA})} = \frac{\alpha_{AA}}{\alpha_{BA}} = \frac{(G_B - G_{BB})}{G_{BA}} = \frac{G_{AB}}{(G_B - G_{AA})} \tag{57}$$

with the elements of the D-matrix on the left hand side only and elements of the G-matrix on the right hand side only. Note that with G homothetic the ratios on the right hand side are indeterminate (0/0). This means that the G-matrix places no constraint on α_{BB}/α_{AB} and α_{AA}/α_{BA} which are then uniquely determined by the D-matrix only.

When we are given the initial conditions

$$\Delta C_A^o = C_A^o - C_{A0} \quad \text{and} \quad \Delta C_B^o = C_B^o - C_{B0} \tag{58}$$

with Fourier transforms $C_A(\beta)$ and $C_B(\beta)$, respectively, then by ternary diffusion theory the α_{ij}'s are determined by

$$\frac{(D_B - D_{AA})}{(D_B - D_A)}\left[\Delta C_A^o - \frac{D_{AB}}{(D_B - D_{AA})}\Delta C_B^o\right] = \alpha_{AA}\int_{-\infty}^{\infty}\bar{C}_A(\beta)\cdot\cos\beta x\cdot d\beta \tag{59}$$

$$\frac{(D_A - D_{BB})}{(D_A - D_B)}\left[\Delta C_B^o - \frac{D_{BA}}{(D_A - D_{BB})}\Delta C_A^o\right] = \alpha_{BB}\int_{-\infty}^{\infty}\bar{C}_B(\beta)\cdot\cos\beta x\cdot d\beta \tag{60}$$

$$\frac{(D_A - D_{BB})}{(D_A - D_B)}\frac{(D_{AB})}{(D_B - D_{AA})}\left[\Delta C_B^o - \frac{D_{BA}}{(D_A - D_{BB})}\Delta C_A^o\right] = \alpha_{AB}\int_{-\infty}^{\infty}\bar{C}_B(\beta)\cdot\cos\beta x\cdot d\beta \tag{61}$$

$$\frac{(D_B - D_{AA})}{(D_B - D_A)}\frac{(D_{BA})}{(D_A - D_{BB})}\left[\Delta C_A^o - \frac{D_{AB}}{(D_A - D_{AA})}\Delta C_B^o\right] = \alpha_{BA}\int_{-\infty}^{\infty}\bar{C}_A(\beta)\cdot\cos\beta x\cdot d\beta \tag{62}$$

and these complete the closed solution. It is a consequence of the homothetic character of G (cf. Eqs. 56 and 57) that only the D-matrix elements appear within the α-matrix. This greatly simplifies the examination of the evolution on the isotherm for appropriate initial conditions and relative values of the D_{ik}. Indeed, since a single gradient energy damping factor is common to each component of the composition vector (C_A, C_B) on the isotherm its relative effect on the evolution can be ignored. Accordingly, our earlier arguments based upon a neglect of gradient and strain energy effects is now seen to be valid (Kirkaldy and Purdy; 1). We conclude by paraphrasing that argument.

For clarity we write out solutions (7) in full, viz.,

$$C_A - C_{A0} = \alpha_{AA}\int_{-\infty}^{\infty}\bar{C}_A(\beta)\cdot\exp\left[-(D_A\beta^2 + G\beta^4)t\right]\cos\beta x\,dx$$
$$+\alpha_{AB}\int_{-\infty}^{\infty}\bar{C}_B(\beta)\cdot\exp\left[-(D_B\beta^2 + G\beta^4)t\right]\cos\beta x\,dx \tag{63}$$

$$C_B - C_{B0} = \alpha_{BA}\int_{-\infty}^{\infty}\bar{C}_A(\beta)\cdot\exp\left[-(D_A\beta^2 + G\beta^4)t\right]\cos\beta x\,dx$$
$$+\alpha_{BB}\int_{-\infty}^{\infty}\bar{C}_B(\beta)\cdot\exp\left[-(D_B\beta^2 + G\beta^4)t\right]\cos\beta x\,dx \tag{64}$$

We have shown in Kirkaldy and Purdy (1) that at the critical point the matrix

$$\rho[D_{ik}] = \mu_{AA} \begin{bmatrix} (L_{AA} - L_{AB}/\delta) & -(L_{AA} - L_{AB}/\delta)/\delta \\ (L_{BA} - L_{BB}/\delta) & -(L_{BA} - L_{BB}/\delta)/\delta \end{bmatrix} \tag{65}$$

where δ is the slope of the limiting tie line and add the inference from the well-known eigenvalues of the D-matrix that

$$D_A = D_{AA} + D_{BB} = \text{tr}\,D > 0 \quad ; \quad D_B = \det D/\text{tr}D = 0 \tag{66}$$

On shallow penetration into the spinodal zone D_A is unchanged while det D and D_B became negative indicating a growing vector component. From relations (56) and (65) at the critical point we have

$$\frac{-D_{AA}}{D_{AB}} = \delta = \frac{a_{BB}}{a_{AB}} = \left(\frac{dX_B}{dX_A}\right)_C \tag{67}$$

and from (57) and (65)

$$\frac{D_{BB}}{D_{AB}} = \frac{L_{BA} - L_{BB}/\delta}{L_{AA} - L_{AB}/\delta} = \frac{a_{BA}}{a_{AA}} \tag{68}$$

Assuming that these relations continue smoothly into the shallow spinodal region where

$$D_A \simeq \text{tr}\,D \qquad \text{and} \qquad D_B \simeq \det D/\text{tr}\,D < 0 \tag{69}$$

we conclude from Eqs. (63), (64), (67) and (68) that for any fluctuation of form (58) there is a growing partial composite vector which approximately parallels the limiting tie-line (and its neighbours) and a decaying part which lies obliquely to the tie-lines. Thus all fluctuations will evolve so that irrespective of the detailed kinetics and wave number distribution phase separation will tend towards parallelism with tie-lines adjacent to the test alloy. Superposed upon this will be the usually conjectured trend to wave-number selection such that $-(D_B\,\beta^2 + G\,\beta^4)$ is maximized (Morral and Cahn; 3). Although our analytic demonstration has only a limited generality it is likely that a broader computational exercise would confirm these as general trends.

Acknowledgement

We are grateful to Professor Mats Hillert for advising us on the Becker Method of determining the gradient energy matrix.

References

1. J.S. Kirkaldy, and G.R. Purdy, Can. J. Phys., 47, 865 (1969).
2. D. de Fontaine, Ph.D. Dissertation, Northwestern University, Evanston, Ill (1966).
3. J.E. Morral and J.W. Cahn, Acta Met., 19, 1037 (1971).
4. L.C. Brown, and J.S. Kirkaldy, Trans. AIME, 230, 223 (1964).
5. R. Becker, Zeits. f. Metallunde, 29, 245 (1937).
6. J.S. Kirkaldy, and D.J. Young, Diffusion in the Condensed State, Institute of Metals, p. 229 et seq. (1987).

Liesegang patterned solutions of ternary diffusion equations with precipitate sinks

Y.J.M. Brechet
L.T.P.C.M./E.N.S.E.E.G., Domaine Universitaire, St-Martin d'Hères, Cedex, France

J.S. Kirkaldy
Institute for Materials Research, McMaster University, Hamilton, Ontario, Canada, L8S 4L7

Abstract

Generalizations of Zener's solutions to the diffusion equations in one and two dimensions have been found for the case where dilute precipitation accompanying ternary diffusion goes unstable on account of a negative eigenvalue. In the planar case the eigenfunction corresponding to the negative coefficient D is a periodic Kummerian function of the parabolic coordinate $\lambda = x/\sqrt{t}$. In the important asymptotic limit $\lambda \to \infty$ this solution implies pure spatial periodicity in accord with the Jablczynski scaling relation which is a well-known characteristic of the Liesegang Phenomenon. The cylindrical case exhibits an analogous character but is much richer in pattern, including rings, spirals of multifold character, broken rings and spirals and rotating and expanding versions of the same. All predicted patterns exhibit a univariate or multivariate degeneracy stemming from the physically indefinite initial condition and this calls for the application of a principle of pattern wave number selection. A heuristic principle equivalent to maximizing the path probability is used to remove the main degeneracy thus specifying a unique column or radial wave number

Introduction

The Liesegang phenomenon, which pertains to the evolution of spirals and concentric rings or one-dimensional bands of precipitates in ternary and higher order systems, has been studied for a century (1-3). The most widely quoted theory is due to Wagner (4), as amended by Kahlweit (5,6) and this has been adapted for application to a number of recent experimental studies in materials science (7-10). Also physical chemists and physicists have in the past decade offered new experiments, models and insights (11-17). Often, the spacing Δx_n between bands at a distance x_n from the surface in the one- or two-dimensional experiment can be represented by the geometric series

$$\frac{x_n + \Delta x_n}{x_n} = \frac{x_{n+1}}{x_n} = k \tag{1}$$

and this scaling law is known as the Jablczynski relation (18).

In an earlier short contribution (19) it was remarked that the Wagner theory is in some ways unsatisfactory for it contains a number of "ad hoc" steps required by the discontinuous nucleation and growth process conjectured. Furthermore, its assumptions are in conflict with certain observations on ring zonation in fungi (20) and on Liesegang-banded crystallization in gels wherein the structure proves to be a single dendritic crystal of continuously variable volume fraction. It therefore seemed worthwhile to seek a continuum theory of the process analogous to spinodal decomposition. The framework for such a theory was developed by Kirkaldy (21,22) in a study of internal oxidation based on simultaneous ternary diffusion and precipitation and there it was noted that the constitutional parameters within the effective diffusion coefficients were such that the latter under certain circumstances take negative values. The effects of capillarity abet such an outcome and so a version of this destabilizing effect will be invoked and superposed upon conventional boundary value problems involving diffusion relaxation of Heaviside functions (19) or two-dimensional Dirac δ-like functions. In this context it is not surprising that the Liesegang precipitation phenomenon in planar and cylindrical boundary conditions often exhibits a pulsing front which advances as the square root of the time. Precipitation instabilities are also observed in uniformly supersaturated solutions (17)which are formally similar to spinodal decomposition, (23-25) for as we have noted the effective diffusion coefficient is negative and this simulates uphill diffusion. We have investigated this problem within linear stability theory under the same capillarity conditions as invoked here (26).

It is to be noted from the start that there is a degeneracy in our formulation which is generic to this class of free boundary problems In such systems, where there is an evolving supersaturation, the initial and boundary conditions may be specified formally in the normal way (here as Heaviside functions or two-dimensional δ-like functions (27)) but one always understands that an initial fluctuation spectrum (usually unknown or undefineable) is superposed upon the formal conditions and this serves to initiate the pattern-forming instability. The formal conditions together with the invariants (e.g. the mass species) assure that if a coherent asymptotic pattern actually forms and if one finds analogous pattern-forming solutions, these can be regarded as unique up to one or more wave number degeneracies. Any principle which one uses to remove these degeneracies may be deemed equivalent to the unknown components of the early stage of a fluctuation-dominated spectrum which have been amplified into the macroscopic pattern. Also, equivalently, one can state that the mathematical solutions near $t = 0$, which here will be seen to possess unphysical oscillations, must be represented in the decomposition of the early fluctuation-dominated regime.

The Capillarity Model

We refer to a representative ternary constitution and a diffusion-driven precipitation reaction A+B in C→AB. For conciseness a dilute solution C_A, C_B in contact with stoichiometric AB precipitates subject to the capillarity-dependent solubility product (19)

$$C_A C_B = K \exp\left(\frac{4\sigma V}{RT.r}\right) \tag{2}$$

It is assumed that the molar volume V is the same as that of the matrix. To relate the spherical precipitate radius r to the precipitate mole fraction, p, we assume nucleation site saturation on n available sites per unit volume and obtain

$$p = \frac{4}{3}\pi r^3 n = \frac{4}{3}\pi n\left(\frac{4\sigma V}{RT}\right)^3 / \left(\ell n\,\frac{C_A C_B}{K}\right)^3 \tag{3}$$

Site saturation presumes a fixed number and uniform distribution of strong defects in a crystalline or gel substrate. For the kinetic relations to follow we shall require

$$\frac{1}{2}\,p = -P_A\frac{\partial C_A}{\partial t} - P_B\frac{\partial C_B}{\partial t} \tag{4}$$

where

$$P_A = -\frac{1}{2}\frac{\partial p}{\partial C_A} = \frac{2\pi n}{C_A}\left(\frac{4\sigma V}{RT}\right)^3 / \left(\ell n\,\frac{C_A C_B}{K}\right)^4 > 0 \tag{5}$$

and

$$P_B = -\frac{1}{2}\frac{\partial p}{\partial C_B} = \frac{2\pi n}{C_B}\left(\frac{4\sigma V}{RT}\right)^3 / \left(\ell n\,\frac{C_A C_B}{K}\right)^4 > 0 \tag{6}$$

Assuming an infinitesimal volume fraction of precipitates, equal molar volumes of precipitate and matrix and negligible ternary diffusion interactions we can write the modified one-dimensional Fick-type equations for parallel bands (22)

$$-D_A\frac{\partial^2 C_A}{\partial x^2} + \frac{\partial C_A}{\partial t} = -\frac{\dot{p}}{2} \tag{7}$$

and

$$-D_B\frac{\partial^2 C_B}{\partial x^2} + \frac{\partial C_B}{\partial t} = -\frac{\dot{p}}{2} \tag{8}$$

Combining these with Eq. 4 yields

$$\frac{\partial C_A}{\partial t} = \frac{1-P_B}{1-P_A-P_B} D_A \frac{\partial^2 C_A}{\partial x^2} + \frac{P_B}{1-P_A-P_B} D_B \frac{\partial^2 C_B}{\partial x^2} \tag{9}$$

and

$$\frac{\partial C_B}{\partial t} = \frac{P_A}{1-P_A-P_B} D_A \frac{\partial^2 C_A}{\partial x^2} + \frac{(1-P_A)}{(1-P_A-P_B)} D_B \frac{\partial^2 C_B}{\partial x^2} \tag{10}$$

These have the standard form of the ternary diffusion equations, but through Eqs. 5 and 6 are strongly nonlinear. Note in particular that the determinant of coefficients is

$$|D| = \frac{D_A D_B}{1-P_A-P_B} \tag{11}$$

which in view of the positive signs of P_A and P_B and their composition dependencies in Eqs. 5 and 6 can become negative for an appropriate high density n indicating the onset of spinodal-like instabilities (9). The general computational investigation of these equations presents almost insuperable difficulties because of the singularities in $|D|$ and the indeterminate nature of the initial conditions. We accordingly explore a special case wherein $D_B >> D_A$ and in the reaction zone $C_B >> C_A$ such that via Eqs. 5 and 6 $P_A >> 1 >> P_B$ whence $|D| < 0$,

$$\frac{\partial C_A}{\partial t} \simeq -\frac{D_A}{P_A} \frac{\partial^2 C_A}{\partial x^2} - \frac{P_B}{P_A} D_B \frac{\partial^2 C_B}{\partial x^2} \tag{12}$$

and

$$\frac{\partial C_B}{\partial t} \simeq D_A \frac{\partial^2 C_A}{\partial x^2} - D_B \frac{\partial^2 C_B}{\partial x^2} \tag{13}$$

In the simplest limiting low amplitude case with the above specialization the diagonalized form of these is (25,28)

$$\frac{\partial C_A'}{\partial t} = D \nabla^2 C_A' \quad ; \quad \frac{\partial C_B'}{\partial t} = D_B \nabla^2 C_B' \tag{14}$$

where the final solutions have the form

$$C_A = a_{11} C_A'(D) + a_{12} C_B'(D_B) + a_0 \tag{15}$$

$$C_B = a_{21} C_A'(D) + a_{22} C_B'(D_B) + b_0 \tag{16}$$

and the eigen-values and determinant

$$D = -D_A/\overline{P}_A \ ; \ D_B \ ; \ DD_B < 0 \tag{17}$$

We thus proceed to the solutions of Eqs. 14, which due to local mass conservation and relation (2) must in a monotone way reflect the distribution of the destabilizing precipitates in the ternary diffusion field.

The One-Dimensional Case

We assume a formal macroscopic initial condition of a one-dimensional semi-infinite sample such that C_B is uniform and confined while C_A is held constant at one surface and allowed to penetrate freely under that fixed driving force (truncated Heaviside functions). Physically one must further assume a superposed fluctuation perturbation of the initial condition which abets the growth of the instability. The precipitation front is established near this surface and advances in accord with the requirement that for precipitation a solubility product like C_AC_B be exceeded (D has already been derived consistent with this requirement). The solution within the precipitate zone must be matched to the solution in advance of the precipitate zone for a complete specification of the configuration (21). The precipitate distribution p is to be obtained finally by substitution in Eq. 3.

These formal initial and boundary conditions and the expected A depletion with distance suggest a search for parabolic forms for both of the eigenfunctions C_A', C_B'. For the positive eigenvalue D_B

$$C_B \sim \text{erf } x/2\sqrt{D_B t} \tag{18}$$

and this would also be the case for C_A' if D were positive. For the case of the negative eigenvalue (dropping subscripts) we examine the solutions of

$$D \frac{\partial^2 C}{\partial x^2} = \frac{\partial C}{\partial t} \tag{19}$$

Retaining maximum generality we seek solutions of the form

$$C\left(\frac{x}{\sqrt{Dt\varepsilon}}, \tau\right) = C(\lambda, \tau) \tag{20}$$

where

$$\varepsilon = \text{sign of D and } \tau = \ln t/t_o \tag{21}$$

whence the differential equation becomes

$$\varepsilon \frac{\partial^2 C}{\partial \lambda^2} + \frac{\lambda}{2} \frac{\partial C}{\partial \lambda} = \frac{\partial C}{\partial \tau} \tag{22}$$

We here emphasize that our success in finding closed form solutions stems from the inductive choice of the logarithmic time variable in Eq. (21). This has the same impact on the time dependent problems as did Zener's λ-parameter choice in the stationary problem (27).

The free parameter t_o will ultimately play a part in the scaling problem. Undertaking separation of variables via

$$C = \ell(\tau)\, g(\lambda) \tag{23}$$

we obtain

$$d\,\ell/d\,\tau = \beta\tau \text{ or } \ell = \ell_0 e^{\beta\tau} = \ell_0 e^{\beta \ell n t/t_0} \tag{24}$$

where β must be a pure imaginary (β = iω)so as to avoid divergence as $\tau \to -\infty$ or $+\infty$ which correspond to the initial condition ($t \to 0$) and the final state ($t \to \infty$). The λ-dependence assures us that the macroscopic initial condition appropriate to this formulation is the Heaviside function. The oscillations at the time origin are deemed to be equivalent to the fluctuations in nature's initial condition, neither of which have an influence on the macroscopic solutions as $\lambda \to \infty$.

The λ-dependent equation is

$$\frac{d^2 g}{d\lambda^2} + \frac{\varepsilon\lambda}{2} \frac{dg}{d\lambda} - \varepsilon\beta g = 0 \tag{25}$$

This is a special case of the general confluent hypergeometric equation

$$\frac{d^2 g}{d\lambda^2} + Q(\lambda) \frac{dg}{d\lambda} + R(\lambda)\, g = 0 \tag{26}$$

for which solutions can be constructed in terms of Kummerian functions (29, 30). We are concerned here only with the asymptotic ($\lambda \to \infty$) solution which corresponds to relatively short time and deep penetrations. The result is (30)

$$C \sim \cos[\omega\, \ell n\, (x^2/4D\varepsilon t_0)] \tag{27}$$

Since relation (27) is independent of t the periodicity is stationary in space as usually observed for the Liesegang Phenomenon. To obtain the scaling law we write for one period

$$2\pi = \omega\left[\ell n \frac{x_{m+1}^2}{4D\varepsilon t_0} - \ell n \frac{x_m^2}{4D\varepsilon t_0}\right] = \omega\ \ell n \frac{x_{m+1}^2}{x_m^2} \tag{28}$$

whence

$$\frac{x_{m+1}}{x_m} = e^{\pi/\omega} \quad \text{or} \quad x_{m+1} = e^{\pi m/\omega}\, 2\sqrt{D\varepsilon t_0} \tag{29}$$

which is the well-known Jablczinski relation for the Liesegang phenomenon (18).

Cylindrical Coordinates

The most interesting of the Liesegang Phenomena appear in cylindrical boundary conditions including rings and spiral patterns of precipitates. The differential equation to be solved is

$$D\left[\frac{\partial^2 C}{\partial \rho^2} + \frac{1}{\rho}\frac{\partial C}{\partial \rho} + \frac{1}{\rho^2}\frac{\partial^2 C}{\partial \phi^2}\right] = \frac{\partial C}{\partial t} \tag{30}$$

where D is negative. Again we let ε = sign of D and change to the variables

$$\lambda = \rho/\sqrt{\varepsilon D t}\ ,\ \tau = \ell n\, t/t_0 \ \text{and}\ \phi \tag{31}$$

whence

$$\frac{\partial C}{\partial \tau} = \varepsilon\left\{\frac{\partial^2 C}{\partial \lambda^2} + \left(\frac{1}{\lambda} + \frac{\varepsilon\lambda}{2}\right)\frac{\partial C}{\partial \lambda} + \frac{1}{\lambda^2}\frac{\partial^2 C}{\partial \phi^2}\right\} \tag{32}$$

Our first separation of variables proceeds according to

$$C(\lambda, \phi, \tau) = H(\lambda, \phi)\,\ell(\tau) \tag{33}$$

whence

$$d\ell/d\tau = \beta\ell \tag{34}$$

and

$$\frac{\partial^2 H}{\partial \lambda^2} + \left(\frac{1}{\lambda} + \frac{\varepsilon\lambda}{2}\right)\frac{\partial H}{\partial \lambda} + \frac{1}{\lambda^2}\frac{\partial^2 H}{\partial \phi^2} = \beta H\varepsilon \tag{35}$$

As previously, β must be pure imaginary and the macroscopic initial value of $C(\rho = 0)$ must be a cylindrical δ-like function. For the second separation of variables

$$H(\lambda, \phi) = g(\lambda)\, u(\Phi) \tag{36}$$

which yields

$$\frac{d^2u}{d\phi^2} = -\alpha^2 u \tag{37}$$

and

$$\frac{d^2g}{d\lambda^2} + \left(\frac{1}{\lambda} + \varepsilon\frac{\lambda}{2}\right)\frac{dg}{d\lambda} + \left(-\frac{\alpha^2}{\lambda^2} - \beta\varepsilon\right)g = 0 \tag{38}$$

We have taken $-\alpha^2 < 0$ where α is a real integer n since at fixed λ (or ρ) we seek coherency of the form

$$u(\phi + 2\pi) = u(\phi) \tag{39}$$

Accordingly the solution of Eq. 37 is

$$u(\phi) = u_0 e^{in(\phi - \phi_0)} \tag{40}$$

where the phase factor ϕ_0 corresponds to the arbitrary choice of the ϕ origin.

Eq. 38, like Eq. 25, is a general confluent hypergeometric equation. Again our interest is in the asymptotic ($\lambda \rightarrow \infty$) solution with $\varepsilon = -1$ which is

$$C \sim \cos(n\phi + \omega\, \ell n\, [\rho^2/4D\varepsilon t_0] + \text{constant}) \tag{41}$$

This describes concentric rings ($n = 0$) and single ($n = \pm 1$) or multiple ($|n| > 1$) spirals of either sign which are stationary in time (see Fig. 1) and the Jablczinski scaling law as before

$$\rho_{m+1}/\rho_m = e^{\pi/\omega} \tag{42}$$

As in the one-dimensional case, ω is undetermined, thus identifying a free boundary problem.

The precipitate density function is to be obtained by evaluating C_A and C_B via Eqs. 15 and 16 and the product C_AC_B which is to be substituted in Eq. 3. Now one of the interesting eigen-functions is an error function with a large diffusion length and the other is sinusoidal with a small wavelength ($D_B >> |D_A/P_A|$) so the local product C_AC_B will be a superposition of a first and second harmonic, the latter having the main effect of skewing the distribution. Thus via Eq. 3 it is seen that the wavenumber dependency of p will be identical to that of C_A or C_B.

Wave Number And Pattern Selection

The removal of the degeneracy on the separation constant ω represents a difficult generic problem. In other free boundary problems of this type we have had success in applying a principle of maximum path probability or a corollary stationarity in the dissipation. For brevity we here adopt an equivalent principle of quasi-stationarity for the cooperative interaction between the two eigenfunctions and express this through the local quasi-steady expression for wavelength

$$\eta \simeq \varepsilon D/v \tag{43}$$

where v is the velocity of penetration of the precipitation front. The latter can be estimated as

$$v \simeq \frac{d}{dt}(\sqrt{D_B t}) = \frac{1}{2}\sqrt{\frac{D_B}{t}} \tag{44}$$

where D_B is the positive eigenvalue corresponding to the B component, so

$$\eta \simeq 2\sqrt{\frac{D^2}{D_B}t} \tag{45}$$

This can now be used with $\varepsilon = -1$ to estimate ω via (cf. Eq. 32)

$$x_{n+1} = \exp\left(\frac{\pi n}{\omega}\right)\sqrt{D\varepsilon t_0} \tag{46}$$

or

$$\eta_{n+1} = x_{n+1} - x_n = \exp\left(\frac{\pi n}{\omega}\right)\left(\exp\left(\frac{\pi}{\omega}\right) - 1\right)\sqrt{D\varepsilon t_0} \tag{47}$$

It can be seen that for $n = 0$, t_0 must be the time of formation of the first band or ring. Matching this to n for the first ring at $t = t_0$ as obtained in relation (45) we obtain the estimate

$$\omega \simeq \pi/\ell n\left(1 + \sqrt{\frac{4D\varepsilon}{D_B}}\right) \tag{48}$$

which applies to both the one-dimensional and cylindrical cases.

If we substitute ω into Eq. (29) we obtain

$$\frac{x_{m+1}}{x_m} = e^{\pi/\omega} = 1 + \sqrt{\frac{4D\varepsilon}{D_B}} \tag{49}$$

or

$$\frac{x_{m+1} - x_m}{x_m} \simeq \sqrt{\frac{4D\varepsilon}{D_B}} \tag{50}$$

This Jablczynski relation agrees with a result obtained earlier by semi-quantitative arguments (19) and offers a very convenient form for the planning of an experimental test.

References

1. R.E. Liesegang, Naturwiss, Wochenschr., 11, 353 (1896).
2. E.S. Hedges, Liesegang Rings and Other Periodic Structures, Chapman and Hall, London, 1 (1932).
3. S. Veil, Les Phénomènes Périodiques de la Chimie, Herman et Cie, Paris, (1934).
4. C. Wagner, J. Colloid Sci., 5, 85 (1950).
5. M. Kahlweit, Z. Phys. Chem., 32, 1 (1962).
6. M. Kahlweit, Prog. Chem. Solids, S.2., 134 (1965).
7. R.L. Klueh and W.W. Mullins, Acta Met., 17, 59, 69 (1969).
8. Y.S. Shen, E.J. Zdanuk and R.H. Krock, Met. Trans. 2, 2839 (1971).
9. V.A. VanRooijen, E.N. Van Royen, J. Vrigen and S. Radelaar, Acta Met. 23, 987 (1975).
10. P.C. Patnaik, Ph.D. Thesis, McMaster University, (1984).
11. P. Ortoleva, in Theoretical Chemistry, Vol. IV, Ed. H. Eyring, Academic Press, New York, (1978).
12. G. Venzl and J. Ross, J. Chem. Phys, 77, 1302 (1982).
13. P. Ortoleva, Z. Phys. B - Condensed Matter, 49, 149 (1982).
14. P. Ortoleva, in Chemical Instabilities, Eds. G. Nicolis and F. Baras, NATO ASI Series, D. Reidel, Dordrecht, Holland, (1984).
15. G. Venzl, J. Chem. Phys., 85, 2006 (1986).
16. G.T. Dee, Phys. Rev. Lett., 57, 275 (1986).
17. M.E. LeVan and J. Ross, J. Phys. Chem. 91, 6300 (1987).
18. K. Jablczynski and S. Kobryner, Bull. Soc. Chim., 39, 383 (1926).
19. J.S. Kirkaldy, Scripta Met., 20, 1571 (1986).
20. J.A. Bourret, R.G. Lincoln and B.H. Carpenter, Science, 166, 763 (1969).
21. J.S. Kirkaldy, Can Met. Quart. 8, 35 (1969).
22. J.S. Kirkaldy, In Oxidation of Metals and Alloys, p. 101, ASM, Metals Park, Ohio, (1971).
23. J.W. Cahn, Acta Met., 9, 795 (1961).
24. J.S. Kirkaldy and G.R. Purdy, Can. J. Phys., 47, 865 (1969).
25. J.S. Kirkaldy and D.J. Young, Diffusion in the Condensed State, Institute of Metals, London, (1987).
26. Y. Brechet and J.S. Kirkaldy, J. Chem. Phys., 90, 1499 (1989).
27. C. Zener, J. Appl. Phys., 20, 950 (1949).
28. J.S. Kirkaldy, in Advances in Materials Research, Vol. 4, Ed. H. Herman, John Wiley and Sons Inc., New York, (1970).
29. M. Abramowitz and A. Stegun, Handbook of Mathematical Functions, Dover Publications, New York, (1965).
30. Y. Brechet and J.S. Kirkaldy, in preparation.

The continuum theory of nucleation in multicomponent systems

J.J. Hoyt
Department of Mechanical and Materials Engineering, Washington State University, Pullman, Washington 99164-2920, U.S.A.

Abstract

The Cahn-Hilliard continuum theory of nucleation is extended to multicomponent solutions. A system of nonlinear differential equations is derived whose solutions yield the concentration profiles of a critical nucleus and across a flat interface. It is demonstrated that considerable simplifications arise under the assumption of a regular solution. Particular attention is paid to the case of one ideal component in a ternary system. Computations of the work of formation of a critical nucleus indicate that even small additions of a third element to a binary solution can have a significant effect on the nucleation reaction.

Effects of solid-state diffusion on the incubation time for nucleation in solids

K.C. Russell
Departments of Materials Science and Engineering and Nuclear Engineering, Massachusetts Institute of Technology, Cambridge, Massachusetts 02139, U.S.A.

Abstract

Most nucleation in solids is predicated on the aggregation of single atoms into a critical nucleus, which typically contains from tens to thousands of atoms. Clearly solid state diffusion is required for this accumulation. In addition, regardless of how rapid steady state nucleation might be, a characteristic time, known as an incubation time, is required for the formation of even a single critical nucleus. Clearly, the length of this incubation time depends strongly on the diffusional process which is operative in nucleus formation. In some cases, the operative diffusive process is not at all self evident. This paper demonstrates how the quantum mechanical principle of time reversal symmetry may be utilized to identify the path of activation for nucleation, and therefrom the operative diffusive process and the incubation time. Time reversal symmetry will be utilized to obtain the incubation times for homogeneous and heterogeneous nucleation in solids. By contrast, mathematical derivations of the incubation time are cumbersome at best, have been performed only for extremely simple cases, and are often non-physical.

Introduction

The theory of nucleation of first order phase transformations was developed in the early part of this century, primarily to describe the condensation of liquid droplets from a supersaturated vapor phase. Feder, et al. (1) review these early efforts and show that the various treatments came to the same basic conclusion, namely that the rate of steady state nucleation, J_s, was given by:

$$J_s = Z\beta^* N \exp(-\Delta G^*/kT) \qquad (1)$$

where

Z = Zeldovich non-equilibrium factor.

β^* = rate at which atoms or molecules are added to the critical nucleus

N = number of nucleation sites per unit volume

ΔG^* = Gibbs free energy of forming a critical nucleus

kT = the Boltzmann factor

Figure 1 shows schematically the free energy of cluster formation from a supersaturated medium. The maximum free energy, ΔG^*, is the activation barrier for nucleation.

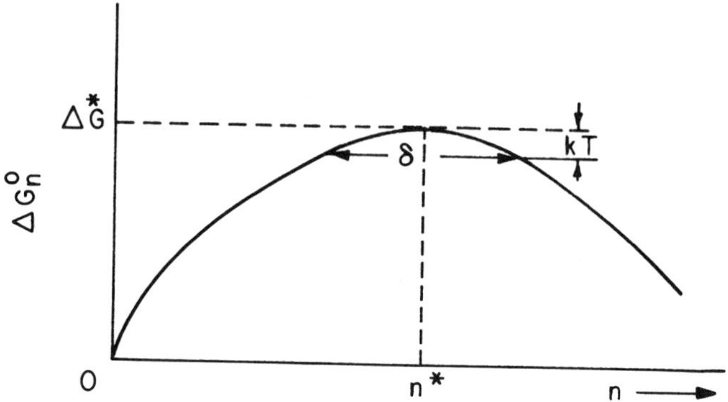

Fig. 1. Schematic representation of activation barrier for nucleation, showing peak height of barrier, ΔG^*, critical nucleus size, n^*, and width of barrier, δ.

Equation 1 gives the rate of formation of critical nuclei per unit volume and time after all initial transients have died out. The steady state rate is expected to obtain until consumption of vapor molecules by growing droplets causes a decrease in the concentration of vapor molecules. The free energy of forming a critical nucleus would then increase and thereby decrease the nucleation rate.

However, steady state nucleation conditions will not obtain immediately on establishment of a supersaturated state. Instead a certain incubation time will be required, during which the steady state concentrations of clusters are built up by aggregation of single atoms or molecules. If the incubation time is long, nucleation may not be observed in a reasonable time, even the steady state nucleation rate is very high. As such calculation of the incubation time has interested many scientists for decades.

Attempts to calculate the incubation time for establishing the steady state rate of condensation from the vapor phase were discussed by Feder, et al. (1). Most attempts were based on approximate solutions of the time dependent equation for the concentration of liquid n-mer in the system, f(n,t). The equation governing f(n,t) is that for diffusional escape of a particle from a potential well.

$$\frac{\partial f(n,t)}{\partial t} = \frac{\partial}{\partial n}\{\beta(n)\, C(n) \frac{\partial f(n,t)/C(n)}{\partial n}\} \qquad (2)$$

Where C(n) is the equilibrium number of n-mer and $\beta(n)$ is the rate at which an n-mer is promoted to (n+1)-mer by capture of single vapor molecules.

Early treatments agreed that the incubation time t, is proportional to $1/\beta^*$, where β^* is the frequency of monomer addition to the critical nucleus. Typically in vapors near the triple point, β^* is the order of 10^{-8} s. Since the critical nucleus is typically no more than a dozen or so molecules, on physical grounds, τ should be the order of 10-100 times this elementary collision frequency. However, various approximations introduced in solving Equation 2 resulted in incubation times ranging from 10^{-8} s to 10^{+12} s. Both extremes are clearly non-physical. Numerical solution of Equation 2 is actually straightforward with the use of modern, powerful computers. Kelton, et al. (2) obtained such a solution, and found that the actual incubation time for the condensation of water vapor near the triple point is the order of 10^{-6} s. However, their calculation assumed that the capture of molecules from the vapor phase is the rate controlling step for nucleus growth. The following sections will show that even in this simple case, deduction of the rate controlling step is not a simple matter.

Time Reversal Symmetry

Feder, et al. (1) utilized the principle of time reversal symmetry to deduce the path of activation, rate controlling step for nucleus formation, and incubation time for condensation from the vapor. Their technique will be outlined, then applied to nucleation in solids.

Time reversal symmetry is a very fundamental principle of physics, which states that microscopic, thermally activated fluctuations form and decay by the same path. Suppose one made a motion picture of such fluctuations forming and decaying, and forgot to mark the beginning and end of the film. By time reversal symmetry, one could not fix the time direction arrow by watching the film. Time reversal symmetry is a consequence of the equations of classical and quantum mechanics being symmetric in time. Every microscopic process and its reverse are thus equally probable. Onsager (3) showed that microscopic fluctuations decay by macroscopic laws, so we may use the well-known laws for dissolution of macroscopic particles to deduce the path of decay for critical nuclei. Then time reversal symmetry may be utilized to deduce the path of activation from the path of nucleus decay.

Figure (1) depicts schematically the activation barrier for nucleation of a particle. In the case of condensation from the vapor, we imagine a droplet at critical nucleus size of n^* molecules, and deduce the path of decay. The droplet will perform a random walk along the essentially flat part of the activation barrier, until falling kT or more below ΔG^*. From the theory of fluctuations, particles of free energy kT or more below the barrier peak will never again achieve the peak. As such, sub-critical nuclei which become smaller than $n^*-\delta/2$ will decay steadily back to monomer. As shown in Figure 1, δ is the width of the activation barrier kT below the maximum. To good approximation, $\delta=1/Z$, where Z = Zeldovich factor (1).

Feder, et al. (1) showed that the time to random walk across δ is typically considerably greater than the time to decay from the edge of the barrier to monomer. Then, from elementary one-dimensional diffusion theory:

$$\tau = \frac{\delta^2}{2\beta^*} \tag{3}$$

Equation 3, derived simply from the principle of time reversal symmetry, is of general applicability in nucleation. One needs first an energetic model to calculate ΔG_n^0. Then, time reversal symmetry is used to determine the path of activation and β^* and δ are calculated. Equation 3 for τ may then be used.

In the case of condensation from the vapor in the presence of a high pressure of inert gas (for example condensation of water vapor in air) the rate of nucleus growth is given by the kinetic gas impingement rate of water molecules on the critical nucleus. The calculation of Kelton, et al. (2) is then correct. In the absence of a large amount of carrier gas, however, loss (or gain) of monomer cools (or heats) the cluster because of the loss (or gain) of the latent heat of condensation. Capture of a molecule then heats the cluster and increases the probability of monomer loss. The rate controlling step becomes the loss of latent heat via thermal fluctuations in the arriving and leaving molecules, and a more complex expression for β^* obtains. The incubation time for condensation of water vapor in the absence of a carrier gas is some two orders of magnitude greater than the μs value calculated. Increasing τ from 1 μs to 100 μs is of little concern in cloud chamber experiments, where expansions are relatively slow. The incubation time may be important in condensation of pure vapors in supersonic nozzles, where expansions are extremely rapid (4).

Those interested in details are referred to Feder, et al. (1) for details. It is sufficient to note here that until their work, no one had realized that latent heat removal from the critical nucleus might be the rate controlling kinetic step in nucleation of liquid from the vapor. The physical and mathematical transparency of time reversal symmetry enabled them for the first time to deduce the correct kinetic factor in nucleation of liquid from the vapor

Nucleus Morphologies in Solids

Nucleation of liquid from the vapor is a relatively simple process as compared to nucleation within crystalline solid phases. Firstly, the critical nucleus in homogeneous condensation is a spherical droplet, and in heterogeneous nucleation on a substrate is a spherical cap with a contact angle which satisfies the Young equation. This simple situation is in sharp contrast to Figure 2 from Johnson, et al. (5), which shows several proposed critical nucleus shapes for nucleation of crystalline β phase within crystalline α phase. These shapes have all been proposed to fulfill the requirement that the β critical nucleus be in metastable equilibrium with the α matrix. The shapes of the critical nuclei thus must satisfy the Gibbs-Wulff construction for the equilibrium form of a particle of fixed volume. Case (a) corresponds to an isotropic α:β interfacial energy, so that the equilibrium form is a sphere. In case (b), the particle and matrix have a low energy interface at the top and bottom, which is preferentially exposed. Case (c) corresponds to nucleation of β particles at an α:α grain boundary. In case c the α:β interfacial energy is isotropic, so that the β nucleus is a smoothly curved lens. The free energy of activation for nucleation, ΔG^*, is proportional to the volume of the critical nucleus. Both faceting and the presence of the grain boundary reduce ΔG^* below that for case (a), so are energetically preferred nucleus morphologies.

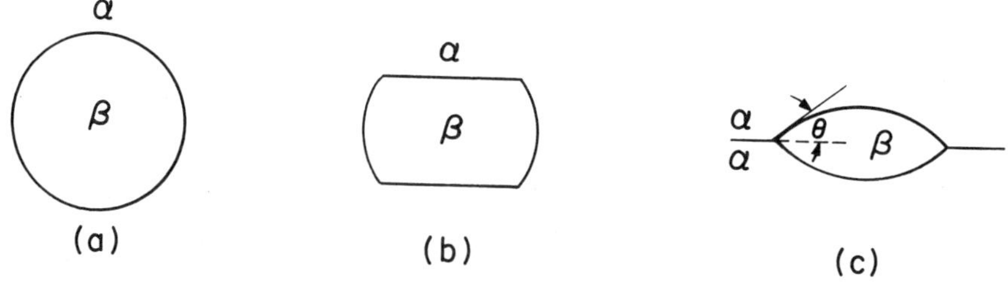

Fig. 2. Proposed nucleus morphologies for nucleation of one crystalline phase inside another. Case a is for homogeneous nucleation with an isotropic interfacial energy; case b is for homogeneous nucleation with an anisotropic interfacial energy; and case c is for nucleation at a grain boundary.

Incubation Times in Solid State Nucleation

The rate controlling step for mass transport to the critical nucleus varies greatly from case to case in nucleation in solids. The spherical β nucleus in figure 2a may be either coherent or incoherent with the α-matrix. In the latter case one might be tempted to identify β^* with the frequency of atomic jumps across the α:β interface. In fact, such jumps, which may occur readily across an incoherent interface, will impoverish the matrix of solute, making further jumps unlikely. The actual rate controlling step is the jump of a solute atom from further out in the matrix to replace the atom which jumped across the interface (6). For case 1a, β^* refers to volume diffusion, whether the interface is coherent or incoherent. Then, τ is given by equation (3), with δ calculated from an energetic model for ΔG_n^o and β^* being the frequency with which solute atoms jump from the matrix into the shell of atoms adjoining the nucleus.

The calculation of β^* and τ is complicated by the presence of facets. If the facet, as in Figure 2b, is immobile and cannot move normal to itself, τ must be calculated on the basis of material addition only at the curved surfaces. Volume diffusion will again control β^*, but the width of the activation barrier, δ, must be calculated on the basis of variations in only the curved surfaces, rather than changes in height as well as diameter of the particle. These issues are addressed by Johnson, et al. (5).

Figure 1-c shows a smoothly curved, lens-like particle lying at an α:α grain boundary. There are two potential paths for solute addition to the particle. Material may add across the curved surface, at a rate controlled by volume diffusion, as in Figure 2a, or by volume diffusion down the α:α boundary to the periphery of the particle, followed by diffusion along the α:β interface to add on the curved surfaces. The latter mechanism was dubbed the "collector plate mechanism" by Aaron and Aaronson (7). The two processes act in parallel, so whichever is faster will dominate. Vander Velde, et al. (8) analyzed the kinetics of these two paths in detail. There are far more addition sites on the curved surface than on the perimeter, but the volume diffusion jump is generally much less frequent than jumps by grain boundary diffusion. In the case of formation of β-phase by a substitutional solute, the collector plate mechanism and grain boundary diffusion are expected to dominate except for the unlikely case of very large critical nuclei forming very near the melting point of the alloy.

The choice of rate controlling diffusive jump is less clear in the case of precipitation of an interstitial solute-rich phase at a grain boundary. This case would include precipitation of Fe_3C particles at grain boundaries in Fe-C alloys. Volume diffusion of the Fe is not required. The diffusion rate of interstitial solutes is generally thought to be only modestly greater at grain boundaries than in the matrix. As such, the greater number of addition sites on the curved surface may more than compensate for the slower rate of volume diffusion. Vander Velde, et al. (8) analyzed the rates of the two paths in detail, and concluded that if $D_b/D_v>500$, D_b is rate controlling, and that if $D_b/D_v<500$, D_v controls. Once the rate controlling step is identified, β^* may be calculated and combined with δ calculated from the energetic model for the nucleus, and τ calculated.

The preceding has shown that use of the principle of time reversal symmetry allows one to unambiguously identify the path of activation and rate controlling step even in complex cases of nucleation in solid phases, and thereby calculate the incubation time, τ.

Experimental Tests of Theory

Experimental measurements of the steady state rate of nucleation in solids are rare; measurements of the incubation time are even rarer. Only two studies come to mind.

Hammel (9) measured both steady-state nucleation rates and incubation times in homogeneous nucleation of a nearly pure SiO_2 phase from a Na_2O-CaO-SiO_2 glass. Russell (6) calculated incubation times on the basis of volume diffusion of SiO_2 (the slowest species) in the matrix and compared his values to Hammel's measurements. The values agreed within about a factor of three over a 25 K temperature range. This agreement is considered to be entirely satisfactory.

Le Goues and Aaronson made a detailed theoretical and experimental study of the homogeneous nucleation kinetics of Co-rich precipitates from dilute Cu-rich, Cu-Co solid solutions (10). Measurements were made over a range of temperatures and Co contents. The Co-rich nuclei are completely coherent with the matrix, so spherical nuclei forming by volume diffusion of cobalt constitutes an appropriate model. Incubation times from 10^2 s to 10^4 s were measured. Agreement was within about a factor of two between theoretical and experimental values if D was taken as defined by the number of vacancies present at equilibrium at the reaction temperature. The excess vacancies retained from the quench from the annealing temperature would relax to equilibrium in only a few seconds; thus,

τ would be controlled by the diffusivity characteristic of the reaction temperature and the agreement between theory and experiment is impressive..

Summary

Calculation of the incubation time for the establishment of steady state nucleation conditions from the classical time dependent nucleation equation is tedious and often misleading. In the simplest case of homogeneous nucleation of liquid from the vapor phase, the rate controlling step for nucleus growth is almost always the dissipation of the latent heat of condensation rather than mass transport to the nucleus. Improperly assuming the latter process to be rate controlling leads to incubation times from roughly 5 to 100 times too small.

The principle of time reversal symmetry may be used to deduce paths of activation and rate controlling process for any case of interest of thermally activated nucleation. In particular the application of this principle allows one to tell whether volume or boundary diffusion is the rate controlling step in homogeneous and heterogeneous nucleation in solids. The two diffusivities often differ by orders of magnitude, so the proper choice is important if an accurate value for τ is to be determined.

There are only two measurements of incubation times in homogeneous nucleation in solids, one being precipitation of SiO_2 from a soda-lime-silica glass, and the other precipitation of Co-rich particles from a Cu-rich solid solution. In both cases the measured values were in good agreement with theoretical values calculated by the principle of time reversal symmetry.

References

1. J. Feder, K. C. Russell, J. Lothe, and G.M. Pound, Advances in Physics 15, 111 (1966).

2. K. F. Kelton, A. L. Greer, and C. V. Thompson, J. Chem. Phys. 79, 6261 (1983)

3. L. Onsager, Phys. Rev. 37, 405 (1931).

4. D. B. Dawson, E. J. Willson, P. G. Hill, and K. C. Russell, J. Chem. Phys. 51, 5389 (1969).

5. W. C. Johnson, C. White, P. Marth, P. Ruf, S. Tuominen, K. D. Wade, K. C. Russell, and H. I. Aaronson, Met. Trans. 6A, 911 (1975).

6. K. C. Russell, Adv. in Coll. Int. Sci., 13, 205 (1980).

7. H. B. Aaron and H. I. Aaronson, Acta Metall. 16, 789 (1968)

8. G. Vander Velde, J. A. Velasco, K. C. Russell, and H. I. Aaronson, Met. Trans. 7A, 1472 (1976).

9. J. J. Hammel, J. Chem. Phys., 40, 2234 (1967).

10 F. K. LeGoues, R. N. Wright, Y. W. Lee, and H. I. Aaronson, Acta Metall., 32, 1865 (1984).

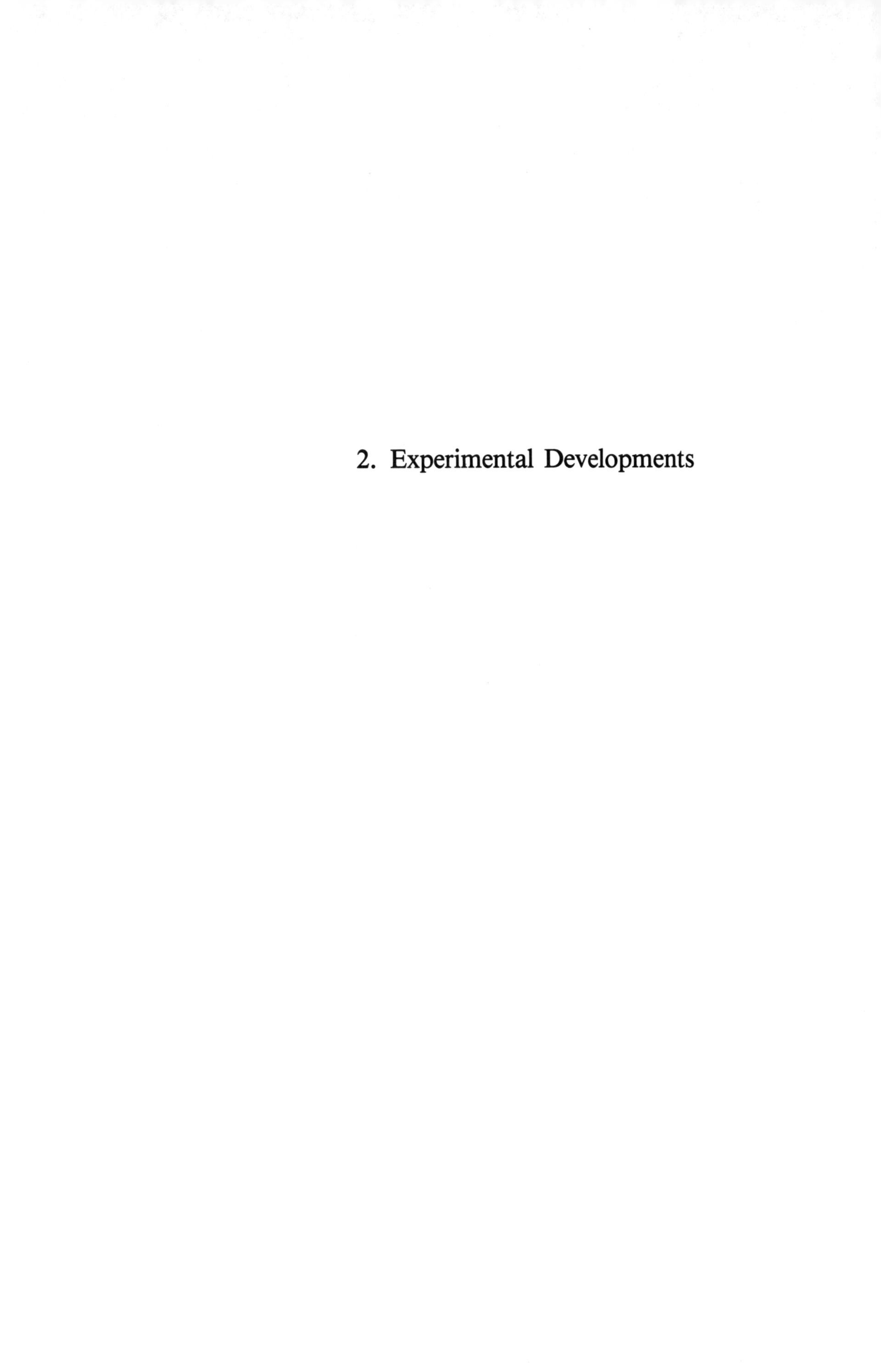

2. Experimental Developments

Diffusion in a two-phase region of the Fe-Cr-Al system

J. Stringer
Electric Power Research Institute, Palo Alto, California, U.S.A.

H.C. Akuezue*
Consultant, Mountain View, California, U.S.A.

Abstract

Diffusion in two-phase systems presents some theoretical problems. In this study, the diffusion in a part of the Fe-Cr-Al system involving the formation of a two-phase region has been investigated. The diffusion path has been determined using electron probe microanalysis and quantitative optical microscopy techniques.

Introduction

In a ternary system, if a diffusion couple is formed between two specimens whose compositions lie in different single phase fields which are separated by a two-phase field, a two-phase region may or may not develop. The latter case corresponds to extremal values of the compositions of the single phase alloys at the interface which are connected by a tie line in the two-phase region. The former case has been reviewed by Kirkaldy and Young (1). They present the possibility of a two-phase region resulting from an interface instability related to a supersaturation which corresponds to a "virtual" diffusion path dipping in and out of the two phase region. They also treat the stability of a planar interface in terms of a perturbation analysis. In this symposium, Dayananda also reports some recent studies on the formation of a two-phase layer in a ternary diffusion system.

In practical systems, the formation of a two-phase region may have significant technical implications. For example, impurity transport may be significantly more rapid along the phase boundaries. We have conducted a study of the aluminizing of iron-chromium alloys using a conventional pack cementation method (2), and, in support of this, a determination of the diffusion coefficient matrix in the single phase alpha phase field of the Fe-Cr-Al system (3). Under certain circumstances, aluminized Fe-20wt%Cr alloys develop a relatively thick two-phase region between an outer single-phase aluminide layer and the alloy. In operation, an aluminized alloy progressively loses the coating, primarily as a result of the loss of aluminum due to repeated spalling and reformation of the protective Al_2O_3 oxide during thermal cycling. If transport of significant impurities such as sulfur inwards or molybdenum out occurs rapidly along the phase boundaries it is possible that this two-phase layer should not be regarded as part of the useful coating thickness. This paper reports on some observations concerning the growth and composition of the two phase layer.

Experimental Method

The method has been described before (2), and will only be briefly discussed here. Several different aluminizing packs were made using mixtures of Fe-50wt%Al and Al powders in an Al_2O_3 pack with an ammonium chloride activator. The pack was blended in a rotating mixer for 4 days to ensure uniformity. Table I lists the compositions of the four packs for which results are reported here.

TABLE I

Pack	Al wt%	Fe-50wt%Al wt%	Al_2O_3 wt%	NH_4Cl wt%
1	20	29	50	1
2	12	37	50	1
3	20	0	77	1
4	9	40	50	1

Assuming that equilibrium is established in the pack at temperature, these compositions are equivalent to 83at%, 73at%, 100at% and 75at%Al respectively. Packs 1 and 3 lie in the Al + θ region in the Fe - Al binary; pack 2 in the $\theta + \eta$ region; and pack 4 is essentially at the η composition.

The Fe-20%Cr was induction melted under argon using high purity starting materials. The ingots were then cut into smaller sections and homogenized in evacuated quartz capsules at 1273 K for two days. Coupons 8 x 8 x 4 mm were cut from the homogenized material, and their surfaces polished to 600 grit. For the aluminizing treatments the specimen was embedded in the pack material in a cylindrical recrystallized alumina crucible with a tightly-fitting lid. After the aluminizing treatment, the specimens were mounted for metallurgical examination.

Where a two-phase region developed, the relative volume fractions of the two phases were determined using the line intercept method. It is recognized that this method is approximate, but for the relatively simple microstructures in this study, the results seem reasonably consistent. Layer thickness was also measured in the optical microscope.

Mass changes as a result of the aluminizing were determined using a standard analytical balance.

The Al and Cr concentrations were measured using electron probe microanalysis (EPMA) on the polished cross-sections. The accelerating voltage was 15 KV, and the sample current was 0.02 mA. For these analyses, the spot size was 3 μm, and point analyses were made at 5 μm spacing. A ZAF computer program was used to convert the measured intensities to weight and atomic percentages using pure element standards. Surface compositions were also measured, taking five counts at ten randomly selected locations on the as-aluminized surface. In the two-phase region, EPMA line traces were taken parallel to the surface. Hopefully, there was local equilibrium established between the phases, so that any plane parallel to the surface had uniform activity of the two constituents, with compositions linked by tie-lines (1). For this analysis, the beam size was 1 μm. However, it should be remembered that in EPMA analysis on cross-sections, the X-rays are emitted from a region bigger than the beam, and with a maximum cross-section somewhat below the surface. Typically, the emitting region is pear-shaped with a maximum cross-section of the order of 2 - 4 μm. If a feature seen in a cross-section has a thickness below the surface less than 5 μm or so, significant errors can result because of contributions from the underlying material. Similarly, contributions from the neighboring phase will result if the phase boundary is within this distance of the beam. In a traverse of the kind used here, the extremal values of composition are the most accurate. It will be seen that these generally are acquired from relatively coarse microscopical features.

Combining the volume fractions of the phases at a specific depth with the compositions for the two phases determines the mean composition of the alloy at that plane. Data of this kind can then be used to determine the boundaries of the two-phase field, the position of the tie-lines, and the location of the diffusion path.

Experimental Results

Figure 1 shows the weight gain versus time for specimens aluminized in pack 2 at temperatures of 1123, 1173, 1223, and 1273 K, plotted as weight gain squared versus time. At least for the two lower temperatures, the results lie on reasonable straight lines, although it should be noted that they do not pass through the origin. Nevertheless, it can be concluded that the results are consistent with a diffusion-controlled process. Table II shows the concentrations of Al and Cr at the surface for the same specimens used in Figure 1 at 1123 and 1273 K. Within the range of error possible in making these measurements, the composition of the surface remains essentially constant. This also would be expected for a process controlled by diffusion in the alloy.

TABLE II

Surface Compositions (at%) as a Function of Time

Aluminizing Time, h	Al	Cr
	1123 K	
2	63	2
4	64	3
4.5	66	2
8	68	6
	1273 K	
0.5	59	10
1	59	9
1.5	60	10
2.5	33	9

The results for the kinetics at the three higher temperatures correspond to an apparent activation energy of 30 kcals/mole, but the rate at 1123 K is not consistent with this.

Figure 2 shows the general appearance of the cross-section for specimens aluminized in packs 2 (A), 1 (B), and 3 (C) for 5 h at 1173 K. There is a single phase layer adjacent to the alloy, and a two-phase layer outside this. In the case of the specimen aluminized in pack 2 there appears to be a thin outer single layer, but examination over a wider field makes this less certain. Figure 3 shows specimens aluminized in pack 2 at 1123 and 1223 K. It is clear that as the temperature increases an outer single phase layer appears and the two-phase layer diminishes in thickness: in some cases, the two-phase layer was virtually absent, and a planar interface between the single-phase regions appeared to be stable at the higher temperature. The inner single phase layer grows at a parabolic rate, again consistent with diffusion control, but the two phase layer does not, except at the lowest temperature.

Careful examination of Figure 2(a) shows that there appears to be continuity between the inner single phase layer and the outer single phase layer (notice particularly the continuous path in the center of the field). The situation in Figure 2(b) and 2(c) is much less clear. The inner single-phase layer appears to be dark etching, but nevertheless appears to be continuous with the lighter etching phase in the two phase layer. This is clearer in Figure 3(a). In Figure 3(b) there appear to be three distinct layers, with the irregular interface between the outer two phases.

Figure 4 shows the composition profiles through the aluminized layer for a specimen aluminized in pack 2 for 5 h at 1173 K: these are the same conditions as for the specimen in Figure 2(a). The fluctuations correspond to the two-phase region. The outer layer appears to be 73 at%Al, 2 at%Cr, and 25 at%Fe (by difference): this corresponds to θ, (Fe, Cr)Al_3, which is essentially the same as the pack composition. The inner layer at what appears to be end of the two-phase region has a composition close to 53 at%Al, 17 at%Cr, and 30 at%Fe. The Fe-Cr-Al phase diagram is, unfortunately, not well-defined in this region, but (as will be seen) it is difficult to rationalize these two compositions as lying in the same phase field, in spite of the metallographic evidence.

Figure 5 shows the appearance of a specimen aluminized in pack 4 for 5 h at 1173 K. Again, there is an inner single phase layer, a two-phase layer, and some indications of an outer single phase layer. Once again, the metallographic section shows an etching difference between the material of the outer layer and the innermost layer, but no discernible phase boundary: the second phase in the two-phase layer has clear boundaries with both. The two-phase layer is more-or-less columnar, with what appear to be isolated precipitates towards the outer surface: this is described by Kirkaldy and Young (1) (theorem 15) as the structure expected when a diffusion path passes into a two-phase region from a single-phase one at an angle to the tie lines but exits into another single-phase region. However, they also remark that the columnar precipitates are "rooted in the parent phase". While this certainly would be expected, it does not appear to be the situation in these specimens.

Figure 6 shows the concentrations of Al and Cr determined from EPMA traverses parallel to the interface in the specimen shown in Figure 5. As indicated above, the extremal values of the compositions will be the closest to the actual values. Figure 7 shows the Al and Cr compositions of both phases as a function of depth measured from the apparent interface between the two-phase layer and the outer single phase layer, and the volume fraction of the light-etching phase, which we have tentatively identified as θ. Table III lists the compositions as a function of depth through the two-phase layer, derived from these curves by hand smoothing. These compositions should be the extrema of the tie-lines, and using the lever rule, the last column in Table III can be used to construct the diffusion path. The phase boundaries and the diffusion path resulting are shown in Figure 8.

TABLE III

Distance	Light-etching phase (θ)		Dark-etching phase		
μm	at%Al	at%Cr	at%Al	at%Cr	%θ
0	71	3	62	12	86
3.3	71	3	62	14	62
6.6	71	3	62	15	45
9.9	72	3	62	15	37
13.2	72	3	62	16	30
16.5	71	3	61	16	26
19.8	70	3	61	15	27
23.1	69	3	60	15	32
26.4	67	3	59	14	38
29.7	65	4	57	14	42
33.0	63	4	56	13	42
36.3	61	5	55	13	41
39.6	59	5	54	12	38
42.9	57	6	53	11	34
46.2	55	6	51	11	30

Discussion

Surprisingly, the Al-rich corner of the Fe-Cr-Al diagram does not appear to have been determined. The most recent diagram is that of Rivlin and Raynor (4), and the most Al-rich feature is a three-phase region $\alpha\delta + \beta_2 + \gamma$, where $\alpha\delta$ is bcc, β_2 is FeAl, and γ is $CrAl_2$. However, the diagram as shown is inconsistent with the binary Al-Cr diagram (5): this shows two adjacent single-phase γ phase fields at 900°C, γ_2 and γ_3, both with extended stability ranges: the Rivlin and Raynor ternary implies a very restricted stability range for the binary γ phase. Two boundaries of the Rivlin and Raynor ternary phase field are shown in Figure 8, and it can be seen that the $\alpha\delta + \gamma + \beta_2 / \gamma + \beta_2$ boundary is inconsistent with the boundaries of the two-phase field determined here, recalling that the boundary shown is between the two phase region and the single dark-etching phase. In addition to the β_2 phase in the Fe-Al system, there are three phases with higher Al contents (6): ζ, which is approximately $FeAl_2$; η, containing 71.5at% Al; and θ. The ternary extent of these phases has not been determined, but it is surprising that the range of stability of the light-etching phase, which appears to be θ, is as large as is implied by the two-phase field boundary determined here. Further work is required to identify the phases present in this system.

The data from the outermost 10 μm of the two phase layer lie very close to the Al-rich end of the two-phase field, and the diffusion path is not very well defined in this region, probably because the boundary between the two-phase layer and the outer single phase layer is also not very well defined. In addition, the length of the line trace over which the analysis was performed is only a few times the section lengths of the phases in this region. Over the innermost 20 μm or so, the diffusion path is much better defined and it lies essentially along a constant Cr line, with approximately 9.5at% Cr. This line lies close to the line corresponding to that for which the Cr content equals 25% of the Fe content, and coincides with it at the left hand side. This implies that Fe is lost towards the outside of the layer, as might be expected from the exchange between the aluminum chloride and the alloy.

Nevertheless, there are some puzzling aspects in the path. The path does not appear to exit the two phase region at the left hand side into a single phase region, and this is consistent with the clear boundary between the dark-etching phase and the inner single-phase layer in the micrographs. Because of the poor definition at the right-hand side it is entirely possible that the path does exit along a tie-line, as might be expected, and is suggested by the curve of the path towards the aluminum-rich corner. The apparent lack of a boundary between the light-etching phase and the innermost layer might be consistent with the example considered by Kirkaldy and Young (1) where the path enters the two-phase region from a single-phase region and then exits to the same single-phase region. This is illustrated for the Cu-Zn-Sn system in their Figure 11.16, taken from Kirkaldy and Brown (7). However, the shape of the experimental diffusion path, and the inherent consistency in the Fe/Cr ratio towards the inner boundary makes this unlikely. The innermost single- phase layer has etching characteristics which seem significantly different to those of the light-etching phase, and it is possible that the apparent absence of a phase boundary is an etching effect: etching these alloys was difficult, and two separate etchants had to be used to reveal various aspects of the microstructure. It would be expected that the inner single-phase layer would be θ (Fe,Cr)Al, and this does not appear to be one of the constituents of the two-phase region. The boundary between the innermost layer and the two-phase layer appears therefore to correspond to the coexistence of three phases: $\beta + \theta + w$, where w is the dark-etching phase. The existence of such a three-phase triangle does not seem inconsistent with the diagram shown in Figure 8, where the β composition is close to the lower left hand corner. The observed microstructure is consistent with Kirkaldy and Young's analysis (their theorem 17).

In several cases there does not appear to be an outer single phase region. For pack 2, which presumably has an Al activity on the low Al side of θ, this is scarcely surprising, and it is interesting that a two-phase layer involving θ can form: this implies that the presence of chromium increases the stability of θ. However, for it also to be true for pack 3 implies the existence of a depletion zone in the pack bordering the specimen, as was first pointed out by Goward and Boone (8). A particular case for the aluminizing of iron in a pure Al pack similar to pack 3 was examined by Akuezue and Stringer (9), where the external Al concentration attained at 1173 K was approximately 72 at%.

Conclusions

It has been shown that, under some circumstances, a two-phase layer is formed during the pack aluminizing of Fe-20%Cr, and that the two-phase layer may be a substantial part of the aluminized region. Furthermore, the two-phase layer may extend to or close to the outer surface, with little or no outer single phase layer.

Higher temperatures favor the formation of single phase layers with relatively stable planar interfaces. At 1223 K the interface is not entirely planar, but no clear two-phase layer is formed. However, at 1173 K a thick outer two-phase layer is formed in packs ranging from pure Al to 73at% Al, 27at% Fe; lower Al activity packs were not tested in this program. A well-defined outer single phase layer was not formed.

The two phases in the layer appear to be (1) a phase resembling θ $FeAl_3$ and (2) a higher chromium-containing phase, here labelled w, which does not appear to correspond to any of the reported phases for the Fe-Cr-Al system. There appears to be an essentially planar boundary between the two phase layer and an inner single phase layer, which appears to be β(Fe,Cr)Al.

The diffusion path appears to exit the low-aluminum end of the two-phase field into a three-phase triangle, β + θ + w. The existence of the inner stable boundary is consistent with the principles outlined by Kirkaldy and Young. The path probably exits the high-aluminum end of the two-phase region along a tie-line into the single-phase θ field, but this point has not been definitely established, principally because of the absence of a well-defined single-phase outer θ layer.

More work is needed on the form of the ternary diagram in the aluminum-rich corner, and on the identification of the phases.

References

(1) J. S. Kirkaldy and D. J. Young, Diffusion in the Condensed State (The Institute of Metals, London, England, 1987).
(2) H. C. Akuezue Diffusion in Fe-Cr-Al System at 900_0C (Ph.D. Thesis, University of California, Berkeley, 1983).
(3) H. C. Akuezue and J. Stringer, Met. Trans.,Vol. 20A (1989) 2767.
(4) V. G. Rivlin and G. V. Raynor, Int. Met. Rev., No.4 (1980) 139
(5) F. A. Shunk, Constitution of Binary Alloys, Second Supplement, (McGraw-Hill Book Company, New York, New York, 1969) 21.
(6) M. Hansen, Constitution of Binary Alloys, (McGraw-Hill Book Company, New York, New York, 1958) 91.
(7) J. S. Kirkaldy and L. C. Brown, Canad. Met. Q., Vol. 2 (1963) 89.
(8) G. W. Goward and D. H. Boone, Oxid. Met., Vol. 3 (1971) 475.
(9) H. C. Akuezue and J. Stringer, J. Electrochem. Soc., Vol. 135 (1988) 1290.

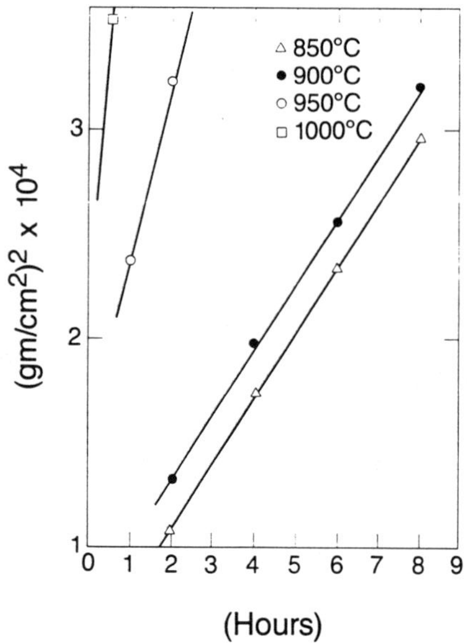

Figure (1) (Weight gain)^2 versus time for Fe-20%Cr specimens aluminized in a pack with a metal composition of 73at% Al, 27at% Fe at 1123, 1173, 1223, and 1273 K.

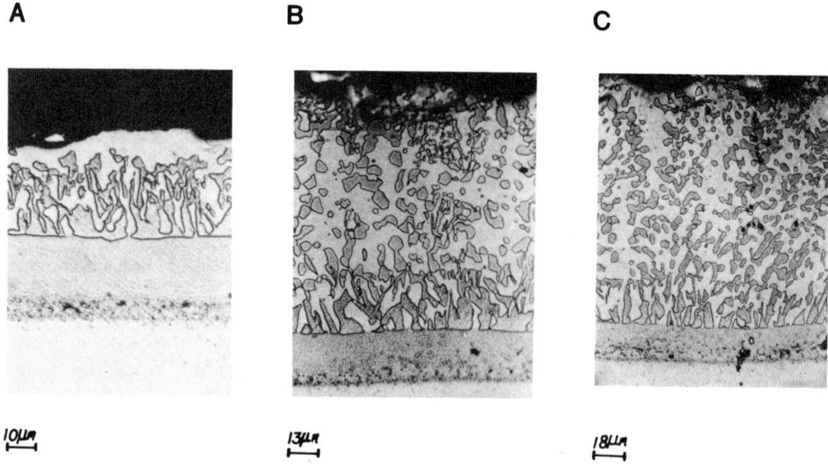

Figure (2) Cross-sections of the aluminized zones in Fe-20%Cr specimens aluminized for 5 h at 1173 K in packs with metal compositions of A - 73at% Al, 27at% Fe; B - 83at% Al, 17at% Fe; and C - 100at% Al.

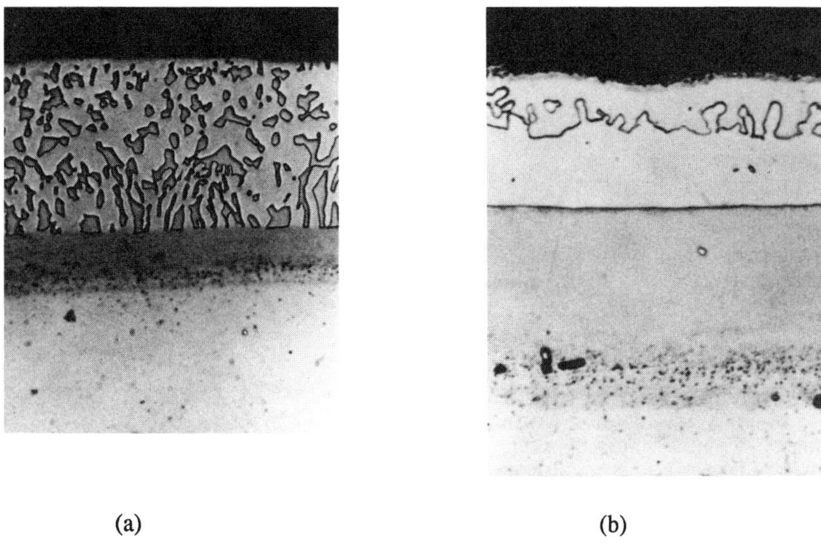

(a) (b)

Figure (3) Cross-sections of the aluminized zones in Fe-20%Cr specimens aluminized for 4 h in a pack with a metal composition of 73at% Al, 27at% Fe at (a) 1123 K; and (b) 1223 K.

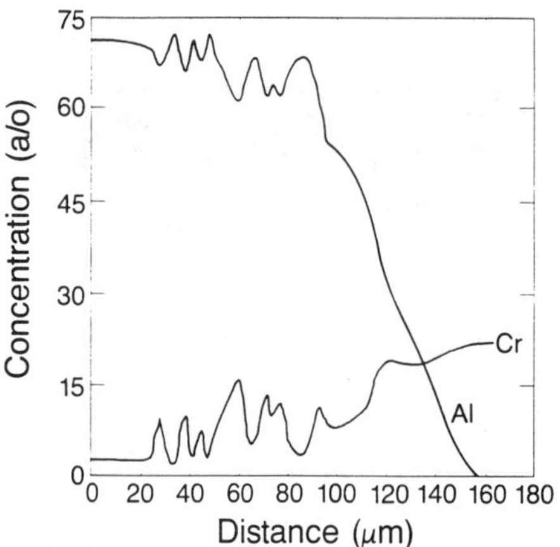

Figure (4) Composition profiles as determined by EPMA through the aluminized zone of a Fe-20%Cr specimen aluminized for 5 h at 1173 K in a pack with a metal composition of 73at% Al, 27at% Fe.

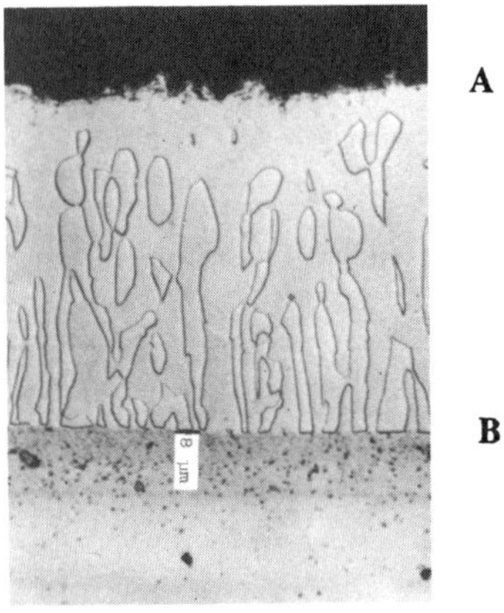

Figure (5) Cross-section of the aluminized region of an Fe-20%Cr specimen aluminized for 5 h at 1173 K in a pack with a metal composition of 75at% Al, 25at% Fe.

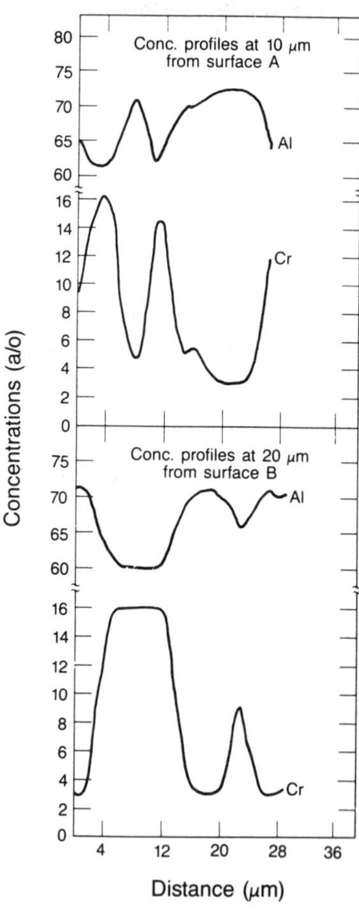

Figure (6) Concentration profiles from EPMA line scans on a cross-section of the specimen shown in Figure 5. The line scans were parallel to the surface at the indicated distances from the outer surface.

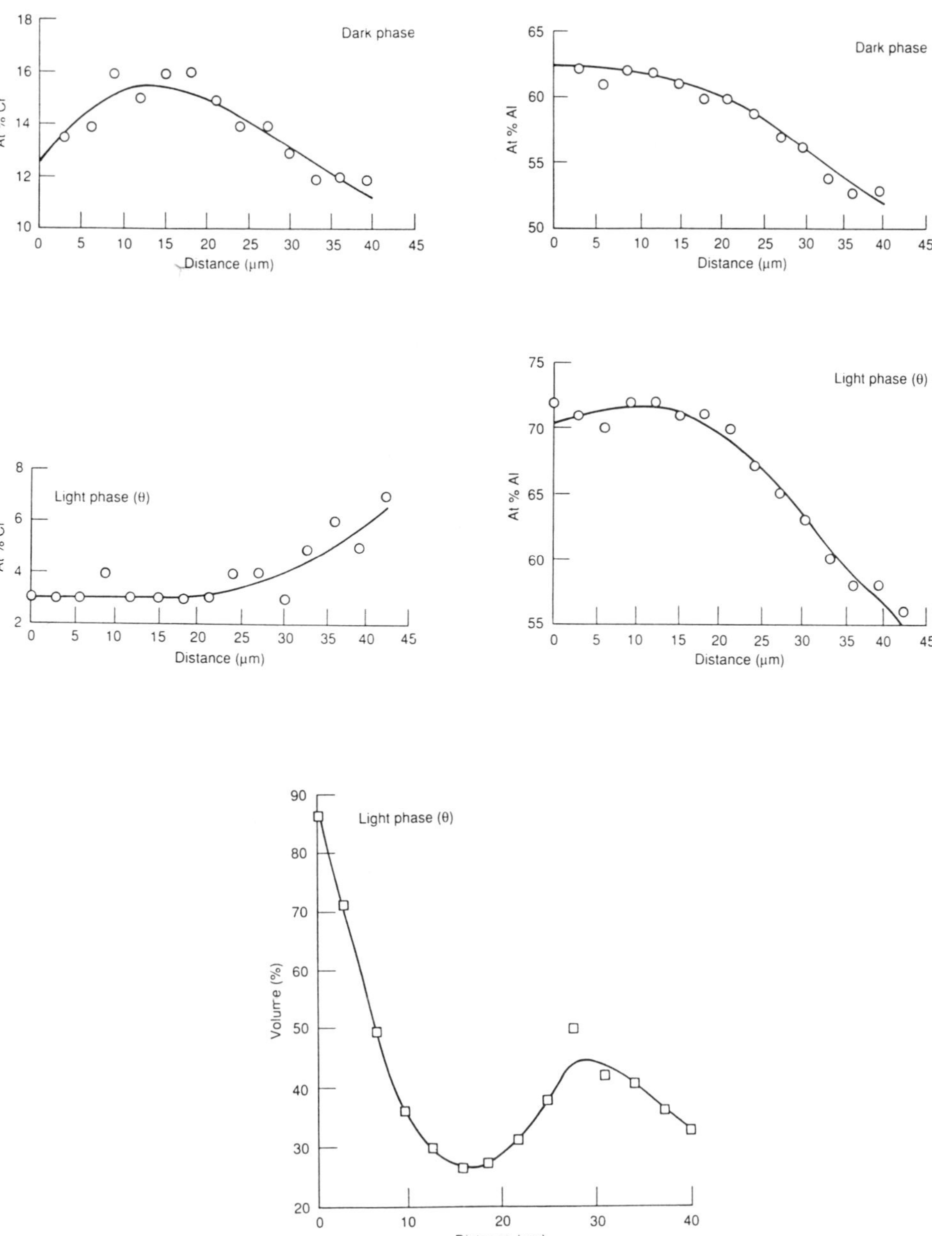

Figure (7) Al and Cr concentration profiles from data similar to that shown in Figure 6 for both phases in the two-phase layer; and the estimated volume fraction of the lighter-etching (θ) phase as a function of depth from the outer surface.

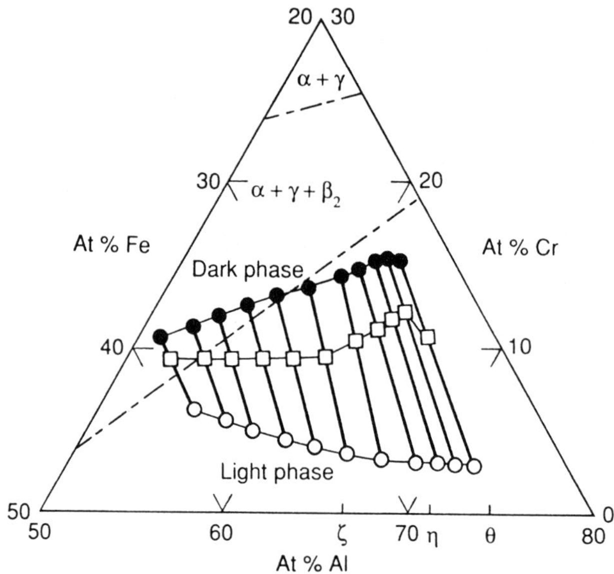

Figure (8) The limits of the two-phase zone and the diffusion path, plotted on a section of the Fe-Cr-Al ternary phase stability diagram at 1173 K. The dashed lines indicate two of the boundaries of a three-phase triangle from the diagram of Rivlin and Raynor (4).

Diffusional transformations in materials with lamellar structures — Case 1: discontinuous coarsening

M. Kaya and R.W. Smith
Queen's University, Department of Metallurgical Engineering, Kingston, Ontario, Canada, K7L 3N6

Abstract

The morphologies and the kinetics of the Pb-Cd, Zn-Cd and the Cu-$MgCu_2$ eutectic alloys have been studied at temperatures 353 to 810 K by scanning and transmission electron microscopy. The fault structure of the initial material was found to effect the final morphology of the coarsened cells. The analysis of the growth kinetics suggests that the reaction is controlled by grain boundary diffusion. The reported data on discontinuous coarsening have been analysed by using Livingston and Cahn's and Fournelle's models.

Introduction

Materials with lamellar structures are commonly used in engineering components. In castings of the multicomponent systems, the last liquid (eutectic) to solidify usually has a rod or lamellar type structure. - Likewise, in some systems, the cooperative growth of two phases side by side may result in a lamellar morphology (eutectoids). It has also been observed that, in other materials, the precipitation of a second phase behind a migrating boundary may result in a lamellar mixture (discontinuous precipitation).

These mixtures of extending solid phases are inherently unstable, particulary at high temperatures where diffusional processes are relatively rapid. Since, these materials usually possess a very high surface area per unit volume, changes occur to produce a progressive reduction in total interfacial area of the system; this is usually manifested as a reduction in the total interfacial area of the various interfaces leading to a change in the initial structure. This may occur by a reaction where the initial fine lamellar mixture is replaced by a coarse product behind the migrating grain boundary (1-20). This reaction has been termed discontinuous coarsening due to its similarities to the discontinuous precipitation reaction. Several analyses have been developed to describe the discontinuous coarsening reactions. The model developed by Livingston and Cahn (1) give the growth velocity as:

$$V = \frac{8C_b^e}{(C_\beta - C_\alpha)} \frac{D_b \delta}{f_\alpha^2 f_\beta^2 \lambda_2^2} \frac{\gamma V_m}{\lambda_1 RT} (1 - r^{-1}) \qquad [1]$$

where C_b^e is the equilibrium boundary composition of the diffusing element, C_β and C_α are the compositions in the α and β phases respectively, D_b is the grain boundary diffusion coefficient, δ is the grain boundary width, γ is the interfacial energy, f_α and f_β are the volume fractions of the α and β phases respectively, λ_1 is the initial lamellar spacing of the alloy, λ_2 is the coarsened spacing of the material, V_m is the molar volume, and r is defined as the coarsening ratio ($r=\lambda_2/\lambda_1$). This equation expresses a relationship between the velocity of the reaction front and the coarsening ratio (r). Fournelle (2) suggested that when the composition of the coarse lamellae differs from that of the fine lamellae, then a chemical energy term should also be incorporated into the growth equation. Therefore, following Petermann and Hornbogen (3), he derived the growth equation as:

$$V = \frac{-8D_b\delta}{RT\lambda_2^2}\left[(1-P)F_o + \frac{2\gamma V_m}{\lambda_2} - \frac{2\gamma V_m}{\lambda_1}\right] \quad [2]$$

where $(1-P)F_o$ is the fraction of the chemical energy available for the process.

Hillert (4) suggested that the examination of the balance of forces at the interface junction would provide valuable information about the driving forces involved in discontinuous coarsening process. However, in this analysis the fractions of the surface tensions carried by the growing coarse lamellae must be known accurately. In principle, if these quantities are known then it would be possible to construct a complete Gibbs energy diagram and find the concentration difference in front of the coarse lamellae that drives the diffusion.

The materials undergoing discontinuous coarsening may be collected into three groups: eutectics(Group I), eutectoids (Group II) and discontinuously precipitated materials (Group III). An examination of this phenemenon in a variety of materials may, therefore, yield more information on the nature of the driving forces involved in diffusional processes and also the nature of the growth process.

In this paper, a critical view will be presented through the examination of the already published data as well as the new experimental results.

Experimental

Pb-Cd and Zn-Cd samples were prepared as described elsewhere (21). The Cu-$MgCu_2$ eutectic (Cu-9.7wt% Mg) was prepared by induction heating in a graphite crucible under an argon atmosphere. The copper metal was melted first and then Mg was plunged into the melt using a graphite holder. After homogenization, the alloy was cast into 5mm rods and subsequently grown in vertical and horizontal Bridgman-type apparati with growth rates between 1.1 to 11.1 $10^{-5} \times$ m-s^{-1}. Samples were heat-treated at between 673 to 810 K and then polished and finally etched with a concentrated nitric acid solution for a short time (1-2 second). Then they were examined in a JEOL T300 scanning electron microscope. For transmission electron microscopy, samples were prepared by ion-beam milling.

Results and Discussion

Initiation of Coarsening

In general, the free energy terms which give rise to the driving force for the discontinuous coarsening process can be stated as:

$$\Delta F_T = \Delta F_s + \Delta F_c + \Delta F_{st}$$

where ΔF_s is the surface free energy and ΔF_c and ΔF_{st} are the chemical and strain free energies respectively. Depending on the system and the experimental conditions, all of these energy terms may be operative.

In binary eutectics and eutectoid materials, the two phases can be stoichiometric compounds, compounds with an extensive solid solubility or solid solutions. In practice, any mixture is possible. If both phases are stoichiometric compounds, then the only driving force will be due to the surface energy term. If non-stoichiometric compounds or solid solutions are involved, then, depending on the extent of the variation of solid solubility with temperature, a chemical energy term should also be considered. It was observed in several eutectics (22, 23) that if the solid solubility of a phase decreases significantly with decreasing temperature , then a precipitation reaction often takes place immediately behind the solidification front. This will reduce any potential supersaturation appreciably and, as a result, the surface energy term would be expected to be the principal influence on the driving force for such cases. However, the presence of the precipitates may affect the initiation of the coarsening process. If the spacing of the initial material is larger, then boundary migration between the lamellae may result in a discontinuous precipitation, causing a fine mixture of lamellae within the initial lamellar structure. It was observed that when the eutectic material is strained the discontinous coarsening rate is increased (19). Since the lamellae orientation to the boundary will be different for each grain with respect to the applied stress, there will be a stress gradient across the boundary which may drive or initiate the discontinuous coarsening reaction. Apart from the external source of strain, misfit strains may exist at the interface. Since the eutectics and the eutectoid materials are produced at high temperatures, there is a considerable opportunity for the associated defects to arrange themselves in low energy configurations. As a result, the misfit is partially relieved by the formation of interface dislocations and steps. Thus, it may be assumed that the ΔF_{st} term is negligible during discontinuous coarsening reactions. Hence, It can be concluded that, in general, the driving force for coarsening reactions in binary eutectics and eutectoids arises mainly from the ΔF_s term.

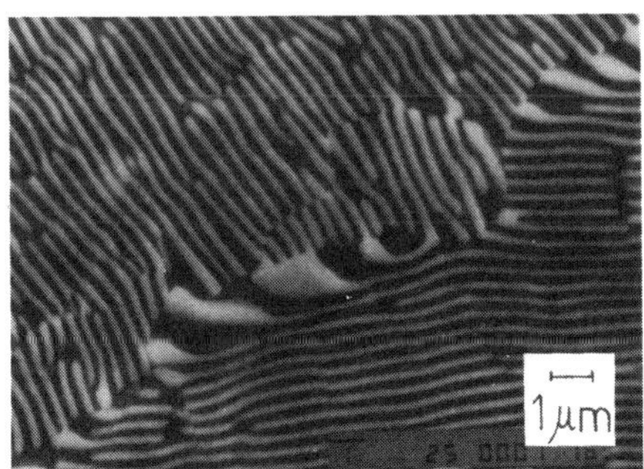

Fig.1. Discountinuous coarsening in the Cu-$MgCu_2$ system heat treated at 800 K for 5h. Note the bending of the coarse lamellae to become normal with the boundary. It is also seen that the fine lamellae have a wavy appearance.

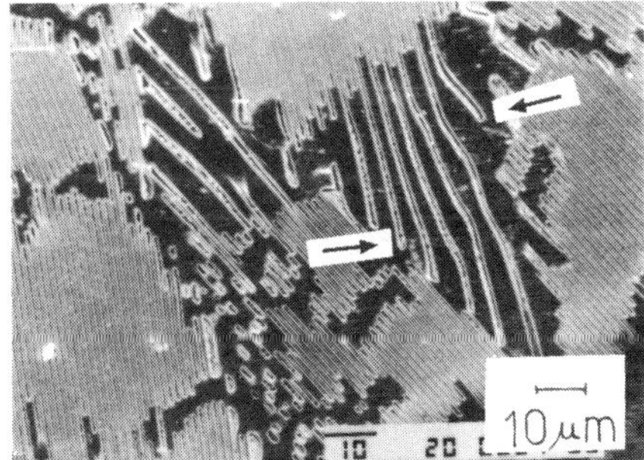

Fig.2. Discountinuous coarsening in the Pb-Cd system. The coarse lamellae grows in the direction of the fine lamellae. Note that the growth of the coarse lamellae ceases when they meet the fault lines (arrowed).

Livingston and Cahn (1) analysed the initiation process; they considered the normal component of the force exerted on the boundary by the lamellae to be:

$$F_{T} = \frac{2\gamma \sin^2\theta}{\lambda_1}$$

where θ is the angle between the lamellae and the grain boundary.

This explains some of the observed morphology where the grain boundary prefers to migrate into a grain whose lamellae makes the largest angle with the boundary. In some cases the coarse lamellae bends itself to become normal to the boundary plane (e.g. Fig.1.) while at other boundaries, the coarse lamellae grows in the same direction as the fine ones (e.g. Fig.2.). In the present study there was no evidence of an effect of lamellar orientation to the boundary on the initiation. Apart from the symmetrically aligned boundaries almost all the boundaries examined were found to be mobile. In the Pb-Cd system, if the initial lamellar spacing of the initial material was large, then the proportion of the boundaries which migrated decreased. For example, only a few of the boundaries were mobile in samples grown at a rate of 1.38 10^{-5} $\times$m-s^{-1}.

Morphology

Two distinct but morphologicaly similar reaction products have been reported in discontinuous coarsening reactions (see Fig.3.). These were termed Type I and Type II by Frebel and Duddek (5). Type I has coarse lamellae nearly parallel to the growth direction of the coarsening cells whereas in Type II coarse lamellae are nearly normal to the growth direction of the cells(see Fig.3.). Both morphologies were observed in the Pb-Cd sytem in the present study.

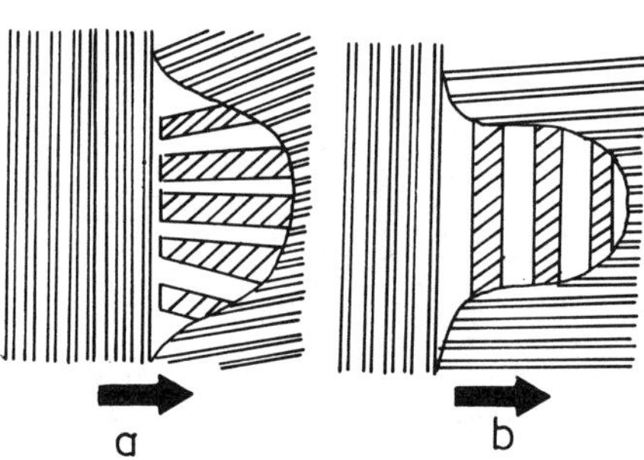

Fig.3. Schematic representation of Type I (a) and Type II (b) morphology. Note that the arrows indicate the apparent growth direction.

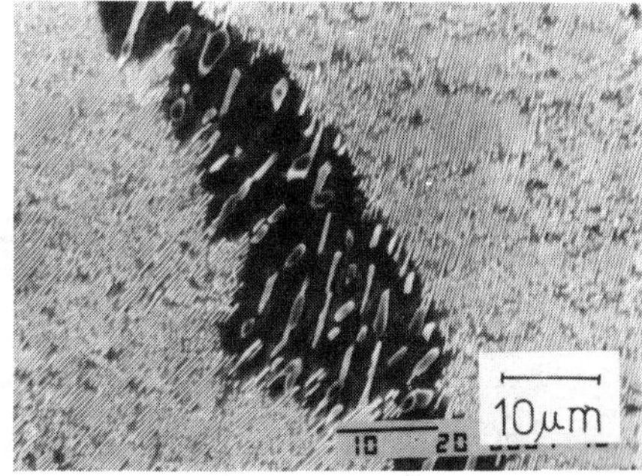

Fig.4. Discountinuous coarsening morphology in the Zn-Cd system. The structure of the cells is quite irregular compared to the Pb-Cd system (Fig.2). Note that the fine lamellae are shorter then in the Pb-Cd system and also they are interrupted by the lamellar faults.

The Type II morphology was not observed in the Zn-Cd and Cu-$MgCu_2$ systems. Although they have similar volume fractions of the phases, the fault structure in the

Zn-Cd system differs from that in the Pb-Cd system. The length of the lamellae between the faults in the Pb-Cd eutectic is greater than in the Zn-Cd system (Fig.4.). The interphase boundary energy in the Zn-Cd system is also higher than that in the Pb-Cd system (24). Therefore, the Zn-Cd eutectic may be more susceptible to changes in the growth variables than the Pb-Cd system. No quantitative analysis is given for fault formation but such an analysis should be related to the structure and the energy of the interfaces. On the other hand, if the structure of the Cu-$MgCu_2$ eutectic is examined, the appearance of the lamellae, which have a wavy character (see Fig.5.), differs from that of both the Pb-Cd and the Zn-Cd eutectics. It looks like as though the structure adjusts itself by the migration of the steps in response to any change in the growth variables. Therefore, in the Zn-Cd system, the lamellae in the boundary area are already broken into several pieces and cannot compete with the lamellae at the other side of the boundary to penetrate into the other grain. Likewise, in the Cu-$MgCu_2$ eutectic constant change in the lamellar direction occurs at both sides of the boundary. Thus the regularity of the eutectic lamellae at the boundary seems to play an important role in the growth of the Type II morphology. This conclusion is also consistent with the observations that Type II morphology has never been reported in Group III materials whose morphology is less regular than that of eutectics and eutectoids.

The Effect of Continuous Coarsening

The apparent effect of continuous coarsening on discontinuous coarsening is to halt the movement of the migrating boundary. As observed in this study, lamellae at the fault lines recede from each other. When the migrating grain boundary encounters the receding fault line, then the migration of the boundary in that segment may stop due to an insufficient supply of solute. In some cases, when the migrating reaction front encounters a fault line, the direction of the coarse lamellae changes. This may be due to the availability of certain steps at the interface. Since the fault lines also introduce a degree of misfit to the system, a step vector may change giving rise to the bending of the lamellae. The effect of continuous coarsening is much clearer at higher temperatures and longer annealing times. However the experiments were confined to regions where the linear growth rate was operative. The coarse lamellae forming behind the migrating boundary, is not continuous due to its frequent encounters with the fault lines. Since these fault lines also breaks up the lamellae into shorter segments in both growth direction and in the direction perpendicular to the growth direction, the fault structure may have an important effect on both the growth and the morphology of the coarsened cells. The irregular morphologies found in the Zn-Cd system may be due to the different fault structures found in this alloy (Fig.4.). If a part of the coarse lamellae throughout the length of the sample meets the fault line, then, because of the short diffusional distances involved, the solute transported in the grain boundary may assist the migration of the side steps. Presumably this process takes place because it is easier assisting the migration of steps than to form new ones, given that both surfaces of the lamella must join at the fault line. Because lamellae have already receded from the fault line, a new surface will be created by the migration of the boundary. Therefore defects will be bounding both sides of the lamellae, as experimentally observed (see Fig.5.). Further migration of the defects in the interface plane occurs, resulting in local bending. Thus regular defect structures like those shown in Fig.5a., are only occasionally observed, while Fig.5b. shows a more commonly occurring one.

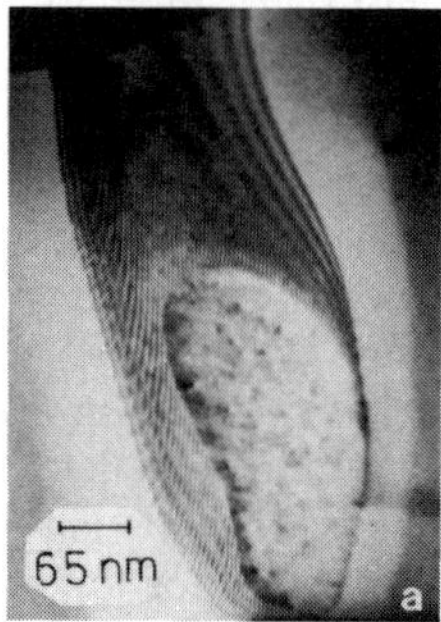

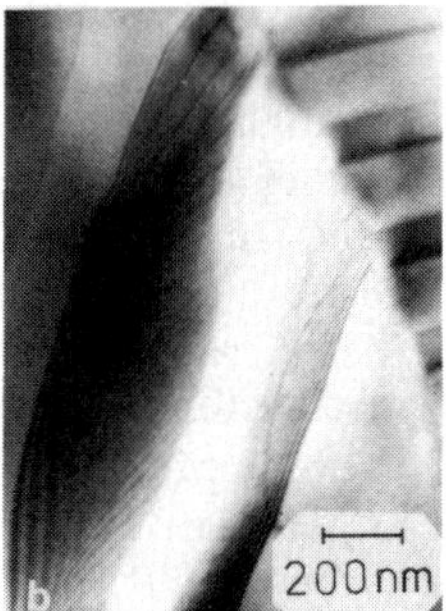

Fig.5. a) The regular defect structure found on the curved section of the coarse lamellae. b) shows a more general view where the migration of the steps at the interface are seen to be occurring.

Kinetics

In order to investigate the kinetics of the reaction, cell widths were measured as a function of the annealing time. The reaction rate was observed to be initially linear with reaction time but to level off. The reaction rate was also found to increase with increasing temperature, as expected, and the rate was particularly large at higher temperatures.

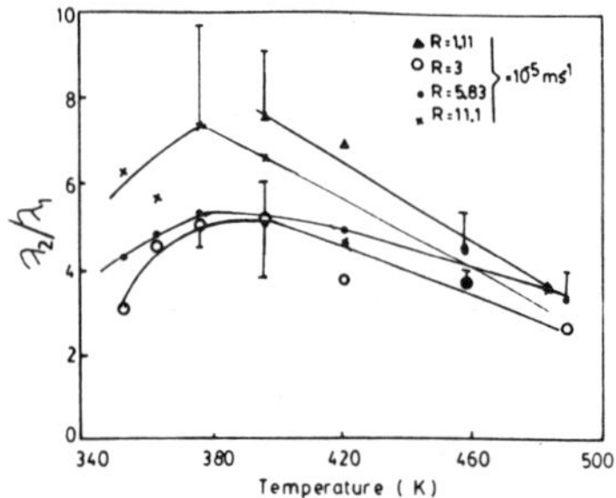

Fig.6. The coarsening ratio vs temperature plot for the Pb-Cd eutectic.

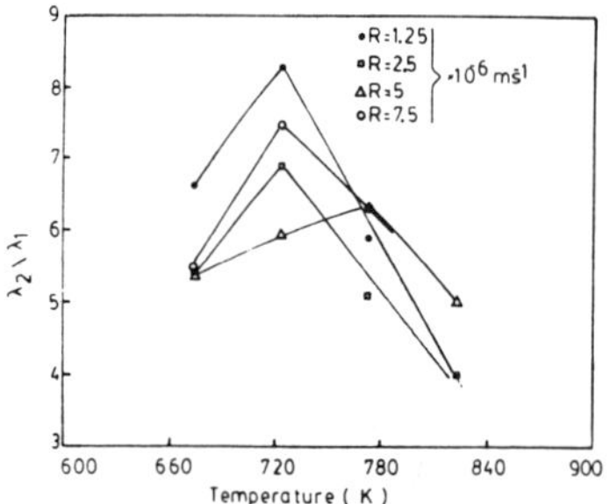

Fig.7. The coarsening ratio vs temperature plot for the Cu-In eutectoid.

Fig.6. shows a plot of λ_2/λ_1 vs temperature. It can be seen that as the temperature increases the lamellar spacing of the cells first increases, passes through a maximum, then decreases again. For the 1.11 10^{-5} $\times$m-s^{-1} samples, no coarsening was detected below 390 K for about 30 days of heat treatment. The maximum in the curves shown in Fig.6. corresponds to $\approx$373 K . For comparison, the lamellar spacing vs temperature data for Group II materials has been extracted from the literature and plotted as shown in Fig.7.. The data reveals behaviour similar to that in the Pb-Cd eutectic. Unfortunately the same plot cannot be made for the Group III materials since all the data reported for these alloys have not been obtained for a constant initial lamellar spacing of the alloy. However, the changes in the coarsening ratio with temperature reported for Group III materials shows a lot of scatter for the different alloy systems. For the Al-29at% Zn alloy, a trend similar to that reported for the Group I and II materials is seen. For the Fe-30wt%Ni-6wt%Ti alloy the coarsening ratio decreases with increasing temperature. On the other hand, for the Cu-15wt%In, Fe-22at%Zn, and Fe-17.6at%Zn, the coarsening ratio remains approximately constant with increasing temperature.

It was mentioned earlier that the first mathematical model of the growth kinetics of discontinuous coarsening reactions was derived by Livingston and Cahn, for the case of eutectoids.

Since the equilibrium boundary concentration (C_ξ) is not known, it can be assumed to be: $C_\xi = kC_o$ where k is a constant and C_o is the average concentration of the solute in the alloy. The $k\delta D_b$ values were calculated from equation (1) and plotted vs 1/T.

Although the application of the Livingston and Cahn analysis to the Cu-In eutectoid (5) has been reported, for comparison, the $k\delta D_b$ values were calculated individually for each growth rate and plotted logarithmically vs 1/T. For the Group III materials such as Al-29at% Zn and Cu-15wt% In, the $k\delta D_b$ values were calculated from the given data; for the other systems they were simply extracted from the published data (2, 9, 10, 25). Figs.8., 9 and 10 show these plots. It can be seen that in most cases there is a linear relationship.

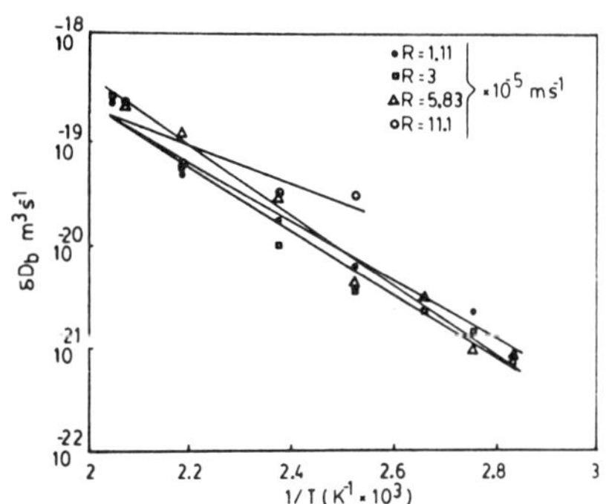

Fig.8. Livingston and Cahn analysis for the Pb-Cd eutectic.

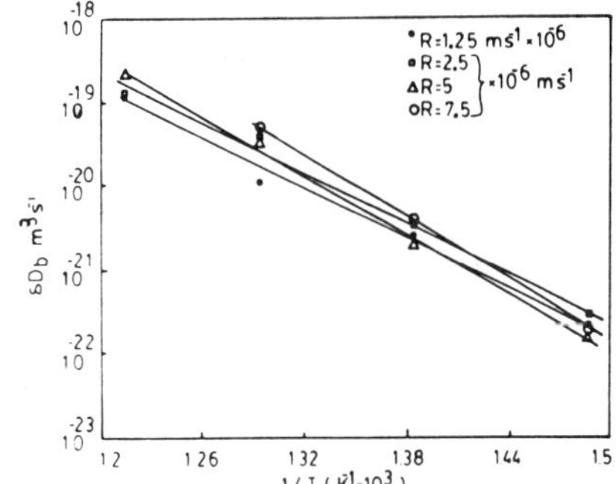

Fig.9. Livingston and Cahn analysis for the Cu-In eutectoid.

Fournelle (2) stated that equations 1 and 2 are functionally the same if P in equation (2) is replaced by unity (i.e. there is no chemical driving force). When this is done, equation 2 reduces to:

$$V = -\frac{16D_b\delta}{RT}\,\frac{\gamma V_m(1-r^{-1})}{\lambda_1\lambda_2^2} \qquad [3]$$

Since the chemical energy term in discontinuous coarsening reactions in eutectic and eutectoid systems should be much less than in the discontinuously precipitated materials, this equation can be tested for the data available on Group I and the Group II materials. Figs.11. and 12 show logarithmic plots obtained for the eutectic and the eutectoid materials respectively. It can be seen that a linear relationship is obtained.

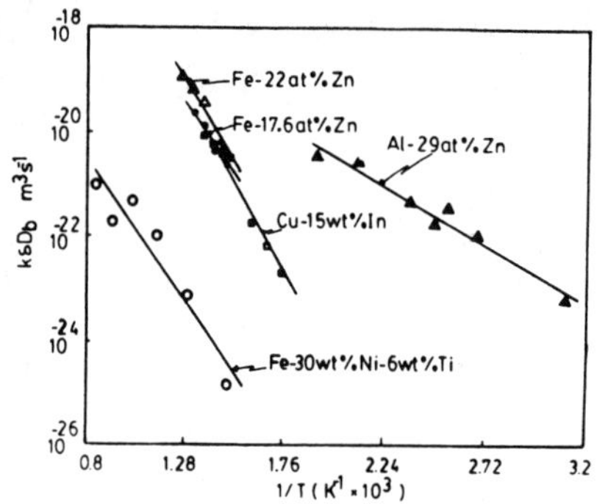

Fig.10. Livingston and Cahn analysis for the Group III materials.

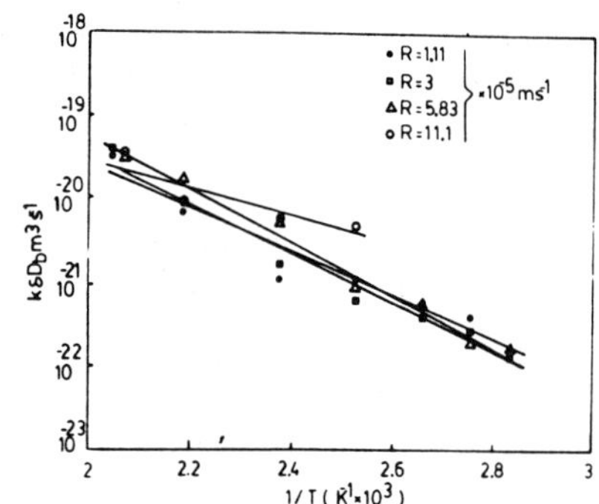

Fig.11. The Petermann and Hornbogen analysis for the Pb-Cd eutectic

Although the activation energy values for the Cd and Pb diffusion in the α/β (Pb-rich and Cd) interfaces are not available, these values can be compared with the data for grain boundary self-diffusion. The activation energies for the grain boundary self-diffusion in Cd and Pb metals are 46-78 and 19-66 kJ-mole^{-1} respectively. Therefore, the values for the grain boundary self-diffusion agree well with the values obtained for the discontinuous coarsening reactions (between 32 and 63 kJ-mole^{-1}). Therefore, the activation energies obtained for the discontinuous coarsening reactions indicate that Pb and/or Cd diffusion through the interphase boundary is the rate controlling step.

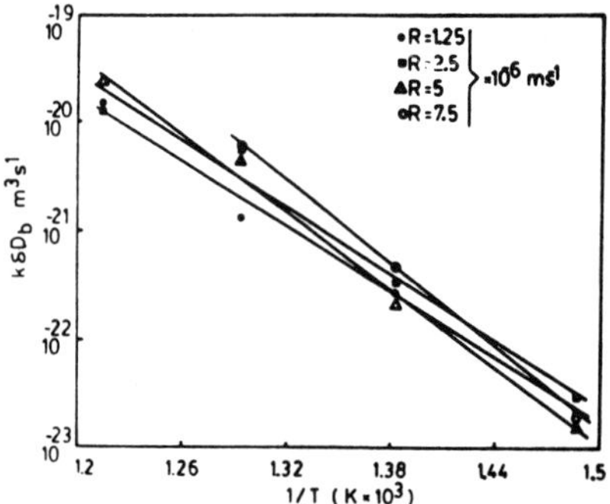

Fig.12. The Petermann and Hornbogen analysis for the Cu-In eutectoid.

If the Group II data are examined, it can be seen that both the Livingston and Cahn and the Fournelle models of cell growth rate give similar activation energies. Since the data, within experimental error, can be fitted reasonably well to both models, it can be said that both models describe the kinetics of the boundary migration for these materials. Frebel and Duddek (5) concluded that since the activation energies found for the discontinuous coarsening reaction in the Cu-In eutectoid are far larger than those for the boundary diffusion in the $\alpha-\alpha$ section, boundary diffusion in the $\alpha-\delta$ and $\delta-\delta$ section of the eutectoid must be the rate controlling process.

Group III materials, relative to Groups I and II, exhibit much more irregular initial morphologies. Furthermore, the effect of the chemical free energy is much more pronounced. The application of the Livingston and Cahn model also yields roughly straight lines. As mentioned earlier, the deviation of the data points at high temperatures for the Livingston and Cahn anaysis was attributed to neglecting the chemical energy term. The magnitude of the chemical energy term decreases as the temperature is raised, due to the continuous precipitation process. Thus, the composition of the initial lamellar product is close to the equilibrium value. Since the quality of the fit for the Livingston and Cahn analysis is good for the low temperature data, where the chemical energy term would be larger, the poorer fit at high temperatures cannot be due solely to the chemical energy term. Although equations [1] and [2] are not too different, the activation energies obtained by the application of the Livingston and Cahn model are smaller than those obtained from the Petermann and Hornbogen kinetics for the Group III materials. However, the application of both models to Group I and II materials yields roughly the same activation energies.

The equation [2] can be rearranged to give;

$$D_b = -\frac{V_2 RT \lambda_2^2}{(1-P_o)F_o + 2\gamma V_m(\frac{1}{\lambda_2} - \frac{1}{\lambda_1})} \qquad [4]$$

It has already been demonstrated that the application of equation [2] to the Group I and II materials with P_o equal to unity resulted in activation energies similar to those obtained when the same data was applied to the Livingston and Cahn model. The first term in the denominator in equation [4] is far larger than the second term and it also decreases with increasing temperature. Therefore, the slopes of the lines will always be larger in this model than in that of Livingston and Cahn. It should also be mentioned that not only should the values of the chemical energy be estimated correctly, these values should be consistent with the temperature change since this will result in a change of the slope. Given that not all the available chemical energy will be used to drive the boundary diffusion, the activation energy values obtained from the two equations will also represent upper and lower limits for the process. To find the second relationship, reliable values for lamellar spacing and cell growth rates have to be known. Group III materials, therefore, would not be ideal for this purpose because the regularity of the cells is much lower than that of Group I and II materials.

The equations (1) and (2) express only one relationship between two unknowns. In order to completely characterize the reaction a second relationship is needed. The application of principles of maximum growth rate or entropy production to the Livingston and Cahn model yields $r \ll 3$. Certain alloys in Group III satisfy this criterion. However, most of the other coarsening ratios experimentally observed were far larger.

Conclusions

Discontinuous coarsening reactions in the Pb-Cd system have been shown to be controlled by grain boundary diffusion. The Livingston and Cahn's model of discontinuous coarsening reactions describes the data obtained on regular lamellar structures such as eutectics and eutectoid materials. Fournelle's model was found to apply to all groups of materials. The cell structures found in the eutectic materials were found to be closely related to the fault structure of the original material.

References

1. J.D. Livingston and J.W. Cahn, Acta Metall., 22 , 495, (1974)
2. R.A. Fournelle, Acta Metall., 22, 1147, (1979).
3. J. Petermann and E. Hornbogen, Z. Metallk., 59 , 814, (1968).
4. M. Hillert, in Phase Transformations, 21 , 415, (1984).
5. M. Frebel and G. Duddek, Mater. Sci. Engng., 32, 17, (1978).
6. R.A. Fournelle and C.P. Ju, in Proc. Int. Conf. on Solid-Solid Phase Transformations, The Metallurgical Society of AIME, p. 957, (1982).
7. R.A. Fournelle, Material Sci. Eng., 63 , 111, (1984).
8. H. Tsubakino and R. Nozato, in Proc. Int. Conf. on Solid-Solid Phase Transformations, The Metallurgical Society of AIME, p. 951, (1982).
9. R.A. Fournelle, Acta Met., 27, 1135, (1979).
10. M. Vijayalakshmi, V. Seetharaman and V.S. Raghunathan, Acta Met., 30, 1147, (1982).
11. Idem, Material Sci. Eng., 52, 249, (1982).
12. K.N. Melton and J.W. Edington, Acta Met., 22 , 1457, (1974).
13. C.P. Ju and R.A. Fournelle, Acta Met., 33 , 471, (1985).
14. S.P. Gupta and G.T. Parthiban, Z. Metallkunde, 76 , 505, (1985).
15. J.S. Brett, G.L. Rehi and E.F. Jarasz, Trans. Met. Soc. AIME, 218 753, (1960).
16. D. Cheetham and N. Ridley, J. Inst. Metals., 99 , 371, (1971).
17. C.W. Spencer and D.J. Mack, in Decomposition of Austenite by Diffusional Processes, Interscience, p. 549, (1962).
18. J.D. Verhoeven, D.P. Mourer and E.D. Gibson, Met. Trans. A, 8A, 1239, (1977).
19. G.D. Delamore, R.W. Van de Merwe, D.P. Dunne and R.W. Smith, in In Situ Composites IV, p. 283, (1982).
20. J.J. Janecek and B.J. PLetka, in Proc. Inter. Conf. on Solid-Solid Phase Transformations, The Metallurgical Society of AIME, p. 963, (1982).
21. M. Kaya and R.W. Smith, Acta Metall. 37, 1657, (1989).
22. D.D. Double and A. Hellawell, J. Cryst. Growth, 6, 10, (1969).
23. R.H. Hopkins and R. Kossowsky, Acta Metall., 19, 203, (1971).
24. L.F. Mondolfo, N.L. Parisi and G.J. Kardys, Mater. Sci. Eng., 68 , 249, (1984-1985)
25. S.P. Gupta, Acta Metall., 34, 1279, (1986).

Selected multiphase diffusion structures in the Cu-Ni-Zn system

M.A. Dayananda and Chun-Li Liu
School of Materials Engineering, Purdue University, West Lafayette, Indiana 47907, U.S.A.

Abstract

Solid-solid diffusion couples assembled with selected α (fcc), β (bcc) and (α + β) two-phase alloys in the Cu-Ni-Zn system were investigated at 775° for the development of diffusion structures and diffusion paths. The microstructures developed in the diffusion zones were studied metallographically and the concentration profiles were determined by SEM-EDAX techniques. The interdiffusion fluxes of the components were calculated directly from the concentration profiles. A diffusion couple with α terminal alloys whose compositions were selected close to the α/(α + β) phase boundary on the Cu-Ni-Zn ternary isotherm developed a complex (α + β) two-phase layer. On the other hand, a multiphase couple assembled with two (α + β) terminal alloys exhibited separation of the phases with the development of adjacent single phase layers of α and β phases. Zero-flux planes for Cu were identified in the diffusion zones of the couples. The diffusion structures for the various couples are described and discussed on the basis of experimental diffusion paths.

1. Introduction

Several multiphase diffusion studies have been carried out to investigate the development of diffusion structures and interfacial instabilities in the Cu-Ni-Zn system with multiphase couples assembled with α (fcc), β (bcc) and γ (cubic) alloys [1-6]. The diffusion structures observed in these studies include the development of both planar and nonplanar α/β and β/γ interfaces depending on the compositions of the terminal alloys. Taylor et al. [1] reported diffusion structures and diffusion paths that varied with time for couples assembled with a β ternary alloy and pure Ni or Cu and annealed at 775°C. Coates and Kirkaldy [2] studied at 775°C the development of morphological instability of α/β interfaces with a series of couples assembled with a binary Cu-Zn α alloy and β ternary alloys of various compositions. A transition form a stable planar to a non-planar α/β interface was observed with increase in the Ni concentration in the β terminal alloy. Sisson and Dayananda [3,4] examined at 775°C diffusion structures developed in diffusion couples assembled with a β Cu-Ni-Zn ternary alloy and a series of α binary alloys of selected compositions and recorded transitions from planar interfaces to non-planar morphologies with compositional changes in the α terminal alloys. They also reported a similar transition from a planar to a non-planar α/β interface for couples assembled with a β terminal alloy and α alloys whose compositions varied along a line of constant Ni/Zn ratio. Wirtz and Dayananda [5] further explored the development of diffusion paths and structures for couples involving the two interfaces, α/β and β/γ, with couples characterized by a common γ alloy joined to a set of α Cu-Ni-Zn alloys; the α/β interfaces exhibited transitions from planar to non-planar and back to planar morphology with decrease in the Cu concentration of the α terminal alloy from 100 to 30 at.pct.

Kim and Dayananda [6] investigated multiphase diffusion in the Cu-Ni-Zn system at 775°C with couples assembled with terminal alloys of similar thermodynamic activity for one of the components in order to study the development of zero-flux planes (ZFP) and flux reversals. ZFPs for the individual components were identified in both single phase as well as multiphase diffusion couples and flux reversals could occur at planar or non-planar interfaces. Also, the ZFP compositions developed in the diffusion couples were close to the compositions of intersection between the diffusion paths and the isoactivity lines drawn through the terminal alloy compositions on a ternary isotherm.

This paper reports recent studies made at 775°C with selected α, β and (α + β) two-phase Cu-Ni-Zn alloys for the development of diffusion structures, diffusion paths and ZFPs. The main objective is to examine the development of a (α + β) two-phase layer in couples assembled with α alloys and to explore the possibility of demixing of phases in couples assembled with (α + β) terminal alloys. The compositions of the alloys employed in the study are indicated on the Cu-Ni-Zn isotherm in Fig. 1 and given in Table I. The single phase α and β

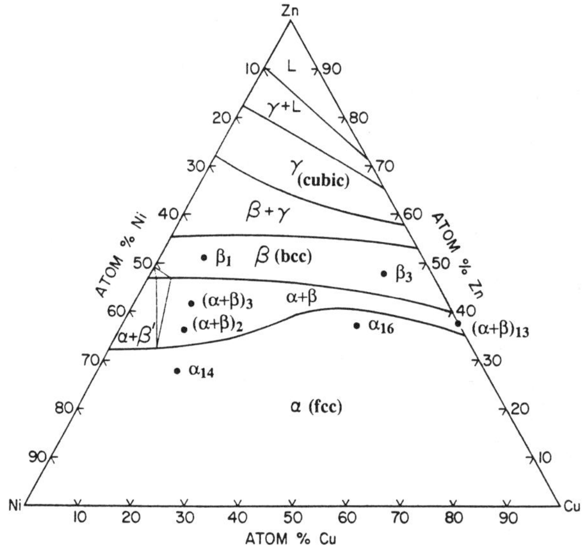

Fig. 1. Compositions of selected α, β and (α+β) alloys on the Cu-Ni-Zn ternary isotherm at 775°C.

TABLE I. Alloy Compositions.

Alloy Designation	Composition*(Atomic percent) Cu	Ni	Zn
α_{14}	15.05	57.50	27.45
α_{16}	43.22	18.91	37.87
β_1	9.04	39.47	51.49
β_3	42.48	9.34	48.17
$(\alpha+\beta)_2$	12.13	52.68	35.19
$(\alpha+\beta)_3$	10.20	48.97	40.83
$(\alpha+\beta)_{13}$	62.50	0.00	37.50

*Average composition is reported for the two-phase alloys.

alloys were selected to have their compositions close to ($\alpha + \beta$) two-phase field, while the two-phase alloys were chosen within the two-phase field. The diffusion structures of all the couples were examined metallographically and analyzed for concentration profiles and diffusion paths by SEM-EDAX analysis. Interdiffusion fluxes were calculated directly from the concentration profiles for the identification of ZFPs.

2. Experimental Procedure

2.1. Preparation of Alloys

The α (fcc) or β (bcc) single phase and ($\alpha + \beta$) two-phase alloys employed in this study were prepared from OFHC copper, electrolytic Ni and high purity Zn by induction melting in alumina crucibles under an argon atmosphere. After melting, the liquid melts were quenched in ice-water. The ingots were machined to remove surface scales, cold-rolled and homogenized for a week at 850°C. The β alloys were very brittle and could not be rolled. However, no gross segregation could be detected in the β alloys after annealing. The alloy ingots were cut into diffusion discs, approximately 1 cm thick, and checked for homogeneity by SEM analysis. The alloy discs were metallographically polished through 0.05 μm alumina and diffusion couples were assembled with discs of selected α or β single phase and ($\alpha + \beta$) two-phase alloys. The discs were clamped together in a Kovar steel jig and the assembled couples were placed in quartz tubes, which were flushed with argon and evacuated to a pressure less than 0.133 Pa and sealed. All the couples were diffusion annealed at 775°C in a Lindberg heavy-duty three-zone tube furnace for 8 hours. The temperature gradient over the length of the capsule was less than 1°C and the temperature controlled to ± 0.5°C. After annealing the capsules were quenched and broken in ice-water in order to retain the high temperature structure of the diffusion zone.

The diffused couples were cold-mounted in quick-mount self-setting resin and cut with an Isomet cut-off wheel to expose sections parallel to the direction of diffusion. The exposed cross-section of each couple was metallographically polished through 0.05 μm alumina and etched with a solution prepared with 2 grams $K_2Cr_2O_7$, 8 ml H_2SO_4, 4 ml saturated NaCl solution, and 100 ml H_2O. The diffusion structures of the various couples were examined metallographically as well as with a JEOL-JSM-35CF scanning electron microscope.

The couples were repolished with 0.3 μm alumina to remove the etched layer and were analyzed for concentration profiles and phase compositions with JEOL-JSM-35CF scanning electron microscope equipped with Tracor Northern Series II energy dispersive X-ray analyzer. The X-ray spectra of the K_α X-ray radiations of Cu, Ni and Zn generated at the energies of 8.047, 7.477 and 8.638 KeV, respectively, were collected and the concentration profiles were determined by point-to-point counting technique. The conversion of X-ray spectra to composition was carried out with a ZAF correction program provided with the Tracor Northern analyzer.

3. Experimental Results and Discussion

The various diffusion couples investigated in this study were assembled with the alloys listed in Table I and the couples presented in this paper are designated by: α_{14}/α_{16}, $\alpha_{16}/(\alpha+\beta)_3$, $\beta_3/(\alpha+\beta)_3$, and $(\alpha+\beta)_2/(\alpha+\beta)_{13}$. These couples consist of one couple with single α phase terminal alloys, two couples where one of the terminal alloys is a single phase α or β alloy and the other terminal alloy is a two-phase (α+β) alloy, and one couple where both terminal alloys are (α+β) two-phase alloys.

3.1. α_{14}/α_{16} couple

The single phase diffusion couple, α_{14}/α_{16}, was assembled with α alloys, where compositions are close to the (α+β) two-phase region, in order to examine the possibility of developing an (α+β) two-phase diffusion layer in the diffusion zone. This expectation was realized for the couple whose diffusion structure presented in Fig. 2 shows a two-phase ($\alpha+\beta$) diffusion layer developed between the two α terminal alloys. This two-phase layer shows an interesting variation in the morphological distribution of α and β phases from one terminal alloy side to the other. Isolated precipitates of the β phase in the α matrix are observed on the side of α_{14}, while an interconnected network of the β phase with islands of α phase develops towards α_{16}. This gradual transition in the phase distribution ends in the formation of a continuous β layer with essentially a planar β/α interface on the side of α_{16}. The diffusion path for the α_{14}/α_{16} couple determined from the concentration profiles is presented on the Cu-Ni-Zn ternary isotherm in Fig. 3. It is apparent that the diffusion path from alloy α_{16} dips into the ($\alpha+\beta$) region at an angle to the equilibrium tie-lines, enters the β single phase region, and returns to the α region crossing the ($\alpha+\beta$) two-phase field along a tie-line.

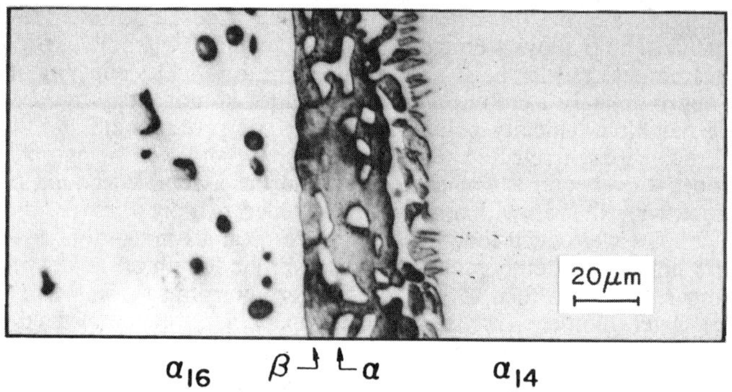

Fig. 2. Diffusion structure showing the development of a two-phase ($\alpha+\beta$) diffusion layer for the couple α_{14}/α_{16} annealed at 775°C for 8 hours.

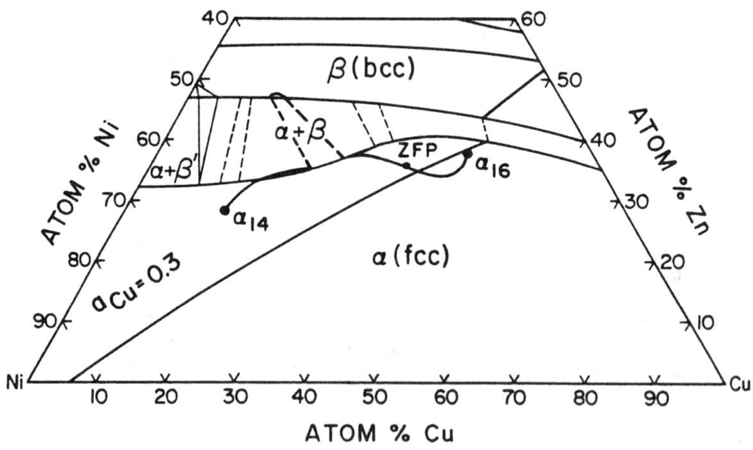

Fig. 3. Experimental diffusion path for the couple α_{14}/α_{16}; a few tie-lines in the ($\alpha+\beta$) two-phase field have been indicated by thin dashed lines.

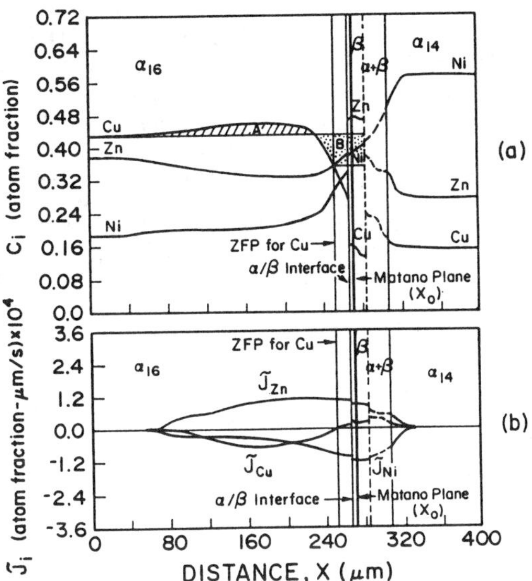

Fig. 4. Profiles of (a) concentrations and (b) interdiffusion fluxes of the various components over the diffusion zone for the couple, α_{14}/α_{16}. A ZFP for Cu developed in alloy α_{16} is identified by the requirement that area A equals area B.

The concentration (C_i) vs distance (x) profiles for the individual components for the couple α_{14}/α_{16} are presented in Fig. 4(a). The parts of these profiles within the ($\alpha+\beta$) diffusion layer are average compositions determined by EDAX analysis and are identified by dashed lines. The original weld-plane of the couple is identified at the Matano plane x_o determined on the basis of the mass balance expressed by

$$\int_{C_i^+}^{C_i^-} (x-x_o)\, dC_i = 0 \tag{1}$$

where C_i^+ and C_i^- refer to the concentrations of the component i in the two terminal alloys of the couple.

For an n-component solid solid diffusion couple, the interdiffusion flux $\tilde{J}_i$ of component i as a function of x can be determined directly from the concentration profile from the relation [7,8]:

$$\tilde{J}_i(x) = \frac{1}{2t} \int_{C_i^+}^{C_i(x)} (x-x_o)\, dC_i \qquad (i=1,2,...,n) \tag{2}$$

The interdiffusion flux profiles for the couple α_{14}/α_{16} calculated on the basis of Eq. (2) are shown in Fig. 4(b). A zero-flux plane (ZFP) where the interdiffusion flux of a component goes to zero is developed by Cu on the α_{16} side of the α/β interface. The location of the ZFP is identified by the requirement

$$\int_{C_i^+ \text{ or } C_i^-}^{C_i(ZFP)} (x-x_o)\, dC_i = 0 \tag{3}$$

Eq. (3) implies that the location of the ZFP for Cu in the couple α_{14}/α_{16} corresponds to the requirement that area A equals area B as shown in Fig. 4(a).

The composition of the ZFP for Cu is identified on the diffusion path in Fig. 3. Also drawn on this figure is an isoactivity line for Cu that passes through the terminal composition point α_{16}. The ZFP composition appears close to the intersection of the diffusion path for the couple and the Cu isoactivity line passing through the terminal α_{16} composition. The proximity of ZFP compositions to the intersections of diffusion paths and isoactivity lines of the individual components drawn through the terminal alloy compositions of the couples have been reported in ternary diffusion studies in the Cu-Ni-Zn and other systems [6,7,9].

3.2. $\alpha_{16}/(\alpha+\beta)_3$ and $\beta_3/(\alpha+\beta)_3$ couples

Two couples, where one of the terminal alloys is a two-phase ($\alpha+\beta$) alloy and the other is a single phase α or β alloy, were investigated. The diffusion paths for the couples, $\alpha_{16}/(\alpha+\beta)_3$ and $\beta_3/(\alpha+\beta)_3$, are presented in Fig. 5. The diffusion structure for the $\alpha_{16}/(\alpha+\beta)_3$ couple is shown in Fig. 6. The α alloy grows at the expense of the ($\alpha+\beta$) two-phase alloy with the dissolution of β phase. In the two-phase diffusion layer identified by TPDL, the β phase islands increase in amount from the α_{16} side towards the $(\alpha+\beta)_3$ side. The TPDL part of the diffusion structure corresponds to the dashed segment of the diffusion path for the couple indicated within the ($\alpha+\beta$) two-phase field in Fig. 5. The path segment cuts across tie-lines over the entire TPDL. Based on the profiles of concentrations and interdiffusion fluxes for the $\alpha_{16}/(\alpha+\beta)_3$ couple one zero-flux plane (ZFP) for Cu is identified within the α region of the diffusion zone. The ZFP composition is close to the intersection of the diffusion path and the Cu isoactivity line passing through the α_{16} composition point, as can be seen in Fig. 5.

For the $\beta_3/(\alpha+\beta)_3$ couple, the β phase grew at the expense of the two-phase alloy with the dissolution of the α phase. The diffusion path for the couple shown in Fig. 5 indicates that the path segment within the ($\alpha+\beta$) region cuts across a few tie-lines at a small angle to them. The two-phase diffusion layer corresponding to this path segment exhibited mostly β phase with very few α islands.

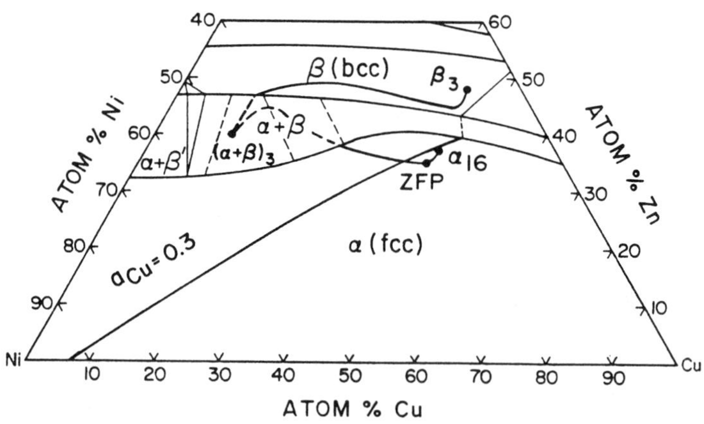

Fig. 5. Experimental diffusion paths for the couples, $\alpha_{16}/(\alpha+\beta)_3$ and $\beta_3/(\alpha+\beta)_3$.

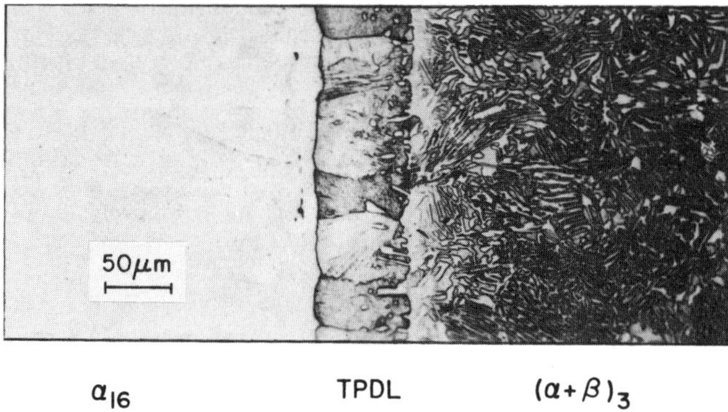

Fig. 6. Diffusion structure developed for the $\alpha_{16}/(\alpha+\beta)_3$ couple annealed at 775°C for 8 hours.

3.3. $(\alpha+\beta)_2$ vs. $(\alpha+\beta)_{13}$ Diffusion Couple

The couple $(\alpha+\beta)_2/(\alpha+\beta)_{13}$ assembled with two $(\alpha+\beta)$ two-phase terminal alloys developed a very interesting diffusion structure with demixing or separation of the α and β phases, as shown in Fig. 7. Adjacent layers of α and β phases with a planar α/β interface formed between the terminal alloys with the α layer about 5 times as wide as the β layer. The β layer grew towards alloy $(\alpha+\beta)_2$, while the α layer grew at the expense of alloy $(\alpha+\beta)_{13}$. In other words, the growth of the β layer resulted from the dissolution of the α phase in the $(\alpha+\beta)_2$ alloy, while the α layer formed by the dissolution of the β phase on the side of alloy $(\alpha+\beta)_{13}$. Kirkendall pores are also visible near the α/β interface on the α side in Fig. 7.

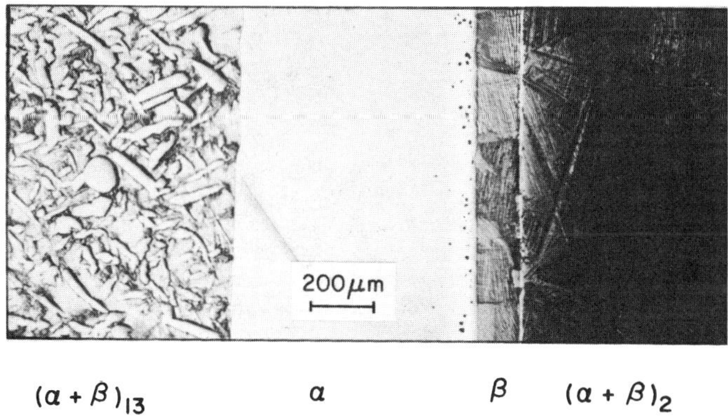

Fig. 7. Diffusion structure for the couple, $(\alpha+\beta)_2/(\alpha+\beta)_{13}$, annealed at 775°C for 8 hours; single phase layers of α and β develop within the diffusion zone.

The diffusion path for the couple is shown in Fig. 8. The path essentially consists of two single phase segments, one in the β phase and the other in the α phase field. The limiting compositions of the α and β segments of the path correspond to the tie-line phase compositions for the $(\alpha+\beta)_2$ and $(\alpha+\beta)_{13}$ alloys and the planar α/β interface formed in the diffusion zone.

The concentration profiles and interdiffusion fluxes for the $(\alpha+\beta)_2/(\alpha+\beta)_{13}$ couple are presented in Fig. 9. Cu concentration profile exhibits a relative maximum and a relative minimum and a zero-flux plane (ZFP) for Cu is observed to develop in the β diffusion layer. Also, the Cu interdiffusion flux shows a flux reversal at the α/β interface. Zn exhibits up-hill diffusion within the α diffusion layer towards the β layer.

3.4. General Comments and Conclusions

One of the important observations of this study is the development of a two-phase (α+β) diffusion layer in a couple assembled with α alloys, such as α_{14}/α_{16}. The two-phase layer is characterized by a gradual variation in the relative amounts of the α and β phases and exhibits morphological variation from isolated β precipitates to a continuous matrix of β phase within its width. The formation of a two-phase diffusion layer in a ternary couple assembled with α (fcc) Cu-Sn-Zn alloys was reported by Kirkaldy and Brown [10]. To understand the development of such layers, he invoked the concept of 'virtual' diffusion path which is a composition path calculated for the couple from a knowledge of ternary interdiffusion coefficients. If the virtual diffusion path for a single phase couple partly falls within a two-phase region, constitutional supersaturation and morphological instability can develop within the couple and lead to the formation of isolated precipitates of a second phase or a two-phase diffusion layer. Couples assembled with α terminal alloys whose compositions are close to the α/(α+β) boundary, as is the case for α_{14}/α_{16}, are candidates for the development of controlled two-phase diffusion structures from single phase alloys.

Another important and significant observation is the opposite phenomenon related to the morphological separation of phases and conversion of two-phase structures to single phase layers with a stable planar interface between them. Such an observation has been reported for

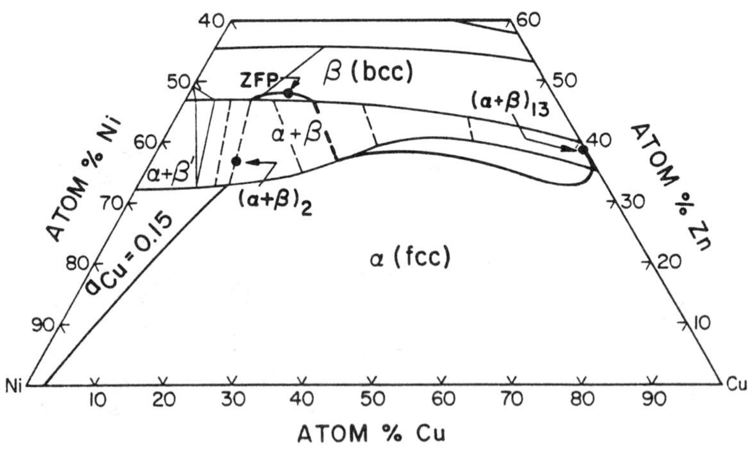

Fig. 8. Experimental diffusion path for the couple, $(\alpha+\beta)_2/(\alpha+\beta)_{13}$.

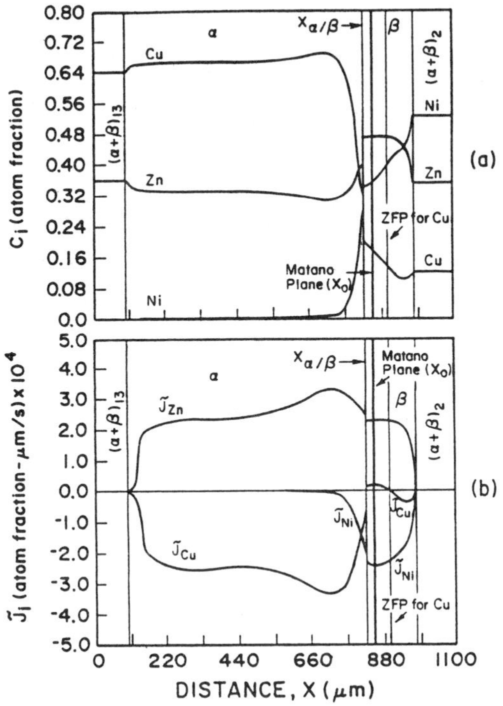

Fig. 9. **Profiles of (a) concentrations and (b) interdiffusion fluxes for the couple $(\alpha+\beta)_2/(\alpha+\beta)_{13}$; Cu develops a ZFP in the β diffusion layer and exhibits a flux reversal at the α/β interface.**

the first time in this study with the aid of the couple $(\alpha+\beta)_2/(\alpha+\beta)_{13}$ assembled with two two-phase terminal alloys. Appreciable elimination or reduction of interfacial boundaries occurs during the process, as each phase becomes unstable in the matrix of the other and gives rise to adjacent single phase layers in the diffusion zone. This study points towards practical applications where a separation of phases into layered structures can be accomplished from two-phase mixtures.

For each of the couples $\alpha_{16}/(\alpha+\beta)_3$ and $\beta_3/(\alpha+\beta)_3$, where only one of the terminal alloys is a two-phase alloy, the main point of interest is the development of a two-phase diffusion layer (TPDL), as the single phase terminal alloy grows at the expense of the two-phase alloy. The TPDL is characterized by variation in compositions as well as relative amounts of the α and β phases over its width, consistent with a two-phase diffusion path cutting across a series of tie-lines. The path segment lying within the two-phase region may be described as part of an S-shaped curve as indicated in Fig. 5 for the couple $\alpha_{16}/(\alpha+\beta)_3$.

The development of ZFPs observed in α_{14}/α_{16}, $\alpha_{16}/(\alpha+\beta)_3$ and $(\alpha+\beta)_2/(\alpha+\beta)_{13}$ indicate the generality of the ZFP phenomenon in couples assembled with single phase and/or two-phase terminal alloys. The ZFP compositions are in the vicinity of the composition points of intersections of the diffusion path and the isoactivity lines passing through the terminal alloys of the couple on the Cu-Ni-Zn ternary isotherm.

References

1. C. W. Taylor, M. A. Dayananda and R. E. Grace, Metall. Trans., 1, 127 (1970).
2. D. E. Coates and J. S. Kirkaldy, Metall. Trans., 2, 2467 (1971).
3. R. D. Sisson, Jr. and M. A. Dayananda, Metall. Trans., 3, 647 (1972).
4. R. D. Sisson, Jr., Ph.D. thesis, Purdue University, 1975.
5. L. E. Wirtz and M. A. Dayananda, Metall. Trans. A, 8A, 567 (1977).
6. C. W. Kim and M. A. Dayananda, Metall. Trans. A, 15A, 649 (1984).
7. M. A. Dayananda and C. W. Kim, Metall. Trans. A, 10A, 1333 (1979).
8. M. A. Dayananda, Metall. Trans. A, 14A, 1851 (1983).
9. C. W. Kim and M. A. Dayananda, Metall. Trans. A, 14A, 857 (1983).
10. J. S. Kirkaldy and L. C. Brown, Can. Met. Quart., 2, 89 (1963).

Calculation of the L_{ij} transport coefficients for $NaCl$-$MgCl_2$-H_2O at 25°C

J.G. Albright
Texas Christian University, Fort Worth, Texas 76129, U.S.A.

D.G. Miller
Lawrence Livermore National Laboratory, Livermore, California 94550, U.S.A.

Abstract

Diffusion coefficients of the system $NaCl$-$MgCl_2$-H_2O have been measured at 25°C over the accessible concentration ranges of both solutes. From these data the phenomenological coefficients, $(L_{ij})_o$ for the solvent-fixed frame of reference have been calculated. Results demonstrate the validity of the Onsager Reciprocal Relations for the mixtures of different valence types.

Introduction

As part of an international collaboration to obtain transport data and thermodynamic data on a representative mixed-valence three-component aqueous electrolyte system, the authors and their colleaques measured diffusion coefficients of the system $NaCl$-$MgCl_2$-H_2O at 25°C over a wide range of concentrations. We denote NaCl as component 1 and $MgCl_2$ as component 2. Measurements were performed at mole ratios C_1:C_2 that were 3:1, 1:1, and 1:3. They were also performed at low concentrations of NaCl in $MgCl_2$ rich solutions and at low concentrations of $MgCl_2$ in NaCl rich solutions. Total concentrations, C_T, ($C_T = C_1 + C_2$) for these experiments were 0.5 M, 1.0 M, 2.0 M, 3.0 M, and 3.8 M (or 3.72 M for the 1:1 ratio and 3.6 M for the 1:3 ratio).

Measurements were made with an optical interferometric diffusiometer that was originally designed and built under the direction of Louis J. Gosting (1). The instrument was used in the Gouy interferometric configuration for all experiments, and in both the Gouy and Rayleigh interferometric configurations for some of the experiments at the 1:1 ratio. With this instrument, diffusion coefficients for ternary systems could be measured with an accuracy that was better than $\pm 1\%$. Results of these measurements have been published for the cases with a 3:1 ratio (2), for all the cases where C_1 was low (3) and for all the cases where C_2 was low (4). The 1:3 case is in press.

Other data published for this system include ternary diffusion measurements by Leaist (5) at low concentrations (0.5 M and lower) with ratios 3:1, 1:1, and 1:3. Tracer diffusion coefficients and viscosities at 25°C have been measured by Mills et al. (6), and conductivities at 25°C have been measured by Bianchi et al. (7). Osmotic coefficients were measured by Rard and Miller over the full range of solubilities (8).

Experimental results

Shown in Figures 1a, 1b, 2a, and 2b are the results of the diffusion measurements. These diffusion coefficients have been measured on the volume-fixed frame of reference based on flow equations written in the form:

$$-J_1 = D_{11}\nabla C_1 + D_{12}\nabla C_2 \qquad (1)$$

$$-J_2 = D_{21}\nabla C_1 + D_{22}\nabla C_2 \qquad (2)$$

In each figure the diffusion coefficients are plotted against the "equivalent" fraction (normality fraction) $x_1 = C_1/(C_1+2C_2)$ for fixed values of the total molar concentration (C_T). Only total concentrations of 0.5, 1.0, 2.0, and 3.0 molarity are included, along with Nernst-Hartley values (9).

Application of the Nernst-Hartley equation gives D_{ij} as both solute concentrations approach infinite dilution for a given ratio of C_1 to C_2 (as defined by x_1). It shows that the D_{ij} at 0.0 M are near linear functions of x_1, as shown in Figures 1a, 1b, 2a and 2b.

A significant feature of these data is that the cross-term diffusion coefficient D_{12} is larger than the other three diffusion coefficients at high total concentrations in NaCl rich systems. This counters the common belief that cross-term diffusion coefficients are small and can be ignored. It also has the implication that NaCl will diffuse backwards against its own concentration gradient when there are opposing gradients of NaCl and $MgCl_2$ which are of equal magnitude and opposite direction.

We see from Figure 1b that the main-term diffusion coefficient of $MgCl_2$, D_{22}, does not vary too greatly from 0.5 to 3.0 total molar concentration for a given value of x_1. In contrast we see that the value of D_{11} for NaCl drops very sharply at low values of x_1 although it remains nearly the same from 0.5 M to 3.0 M at high values of x_1. Thus at low values of x_1 (i.e., solutions rich in $MgCl_2$), the diffusion coefficient D_{11} for NaCl starts higher than the value of D_{22} for $MgCl_2$, decreases substantially as total concentration increases until it is considerably less than D_{22}, and finally approaches half the value of D_{22}. In other words, at low total concentration of $MgCl_2$-rich solutions, NaCl is the fastest diffuser, whereas at high total concentration $MgCl_2$ is the fastest diffuser. This result was completely unexpected.

In order to calculate thermodynamic transport coefficients, it is necessary to have the gradients of chemical potential at each of the solute concentrations at which diffusion coefficients were measured. We used the data of Rard and Miller (8) to calculate these values. Rard and Miller had fit their data to Scatchard's neutral electrolyte equations, and we used their recommended sets of coefficients that were obtained when the fit was to their data only. The necessary supporting activity data for the binary system $NaCl-H_2O$ were those given in an article by Hamer and Wu (10). For $MgCl_2-H_2O$ we used the data given by Rard and Miller (11).

Thermodynamic transport coefficients (Diffusion Onsager Coefficients) were calculated from the chemical potential gradients and the diffusion coefficients calculated for the solvent-fixed frame of reference. These coefficients are defined by the flow equations for the solvent-fixed frame of reference by the equations:

$$-(J_1)_o = (L_{11})_o \nabla\mu_1 + (L_{12})_o \nabla\mu_2 \qquad (3)$$

$$-(J_2)_o = (L_{21})_o \nabla\mu_1 + (L_{22})_o \nabla\mu_2 \qquad (4)$$

The cross-term coefficients should in principle satisfy the Onsager Reciprocal Relations (ORR):

$$(L_{12})_o = (L_{21})_o \qquad (5)$$

Values of $(L_{12})_o$ and $(L_{21})_o$ were calculated for twenty-five sets of mean concentrations at which experiments were performed. In all cases, the cross-term coefficients were equal to each other within four times the propagation of error of an individual coefficient; this represents good agreement of the coefficients.

Values of $(L_{ij})_o$ were also calculated from Leaist's data (5), and these are included in figures given below. In these figures (and in the discussion below) the labels for $(L_{ij})_o$ are written as L_{ij} to simplify notation.

Figure 3 shows values of $L_{11}/(x_1N)$ versus total ionic strength, I, $(I = C_1 + 3C_2)$ of the solutions. Here x_1 is again the equivalent fraction of component 1 and N is the normality (equivalent concentration) of the solution $(N = C_1 + 2C_2)$. We see that for each ratio of $C_1:C_2$ (3:1, 1:1, and 1:3) the values of L_{11} roughly parallel each other and decrease as concentration increases.

A similar graph is shown in Figure 4 for $L_{22}/(x_2N)$ versus ionic strength. Here the values decrease as the ratio $C_1:C_2$ increases at a given ionic strength. The values for the different ratios again generally parallel each other and decrease as the ionic strength increases.

In Figure 5 are given values of $L_{12}/(x_1x_2N)$. Here the values are negative. They also decrease in magnitude as ionic strength increases.

Figure 6a shows $L_{11}/(x_1N)$ versus x_1. In the limit of zero concentration, $L_{11}/(x_1N)$ is nearly a linear function of x_1 that can be calculated from limiting equivalent conductivities of

Na^+, Mg^{2+}, and Cl^- (9). These limiting equivalent conductivities are respectively 50.10×10^{-4} (12), 53.32×10^{-4} (13), and 76.35×10^{-4} (12) m^2 Int. Ω^{-1} equiv.$^{-1}$. The near linearity of $L_{11}/(x_1N)$ is a consequence of the similar values of the limiting equivalent conductivities of Na^+ and Mg^{2+}. As total concentration increases, $L_{11}/(x_1N)$ versus x_1 at fixed total concentration becomes more curved, dropping sharply at high total concentration as x_1 decreases.

In contrast, $L_{22}/(x_2N)$ which is also nearly a linear function of x_1 at infinite dilution, remains nearly linear at each fixed total concentration as total concentration increases. This is shown in Figure 6b.

Finally Figure 7 shows $L_{12}/(x_1x_2N)$ versus x_1. Interestingly this is nearly linear for each total concentration. At infinite dilution $L_{12}/(x_1x_2N)$ will change only slightly, due to the similar values of Mg^{2+} and Na^+ conductivities, as x_1 goes from 0 to 1 (9).

Conclusion

The results of this study show some interesting features. The contrast in the behavior of $L_{11}/(x_1N)$ and $L_{22}/(x_2N)$ each versus x_1 for fixed total concentrations at the progressively higher total concentrations is striking. It is clear that in $MgCl_2$ rich solutions the mobility of Na^+ must become sharply reduced as the concentration of $MgCl_2$ increases.

The results of this study demonstrate the validity of the ORR in concentrated mixed valence electrolyte solutions.

Refinement and analysis of data is still in progress for the 1:1 case. On completion of this analysis, we anticipate publication of values of the Onsager coefficients and the verification of the ORR. As transference number data become available, it will become possible to calculate the ionic transport coefficients l_{ij} (9) for the $NaCl$-$MgCl_2$-H_2O system.

Acknowledgments

JGA expresses gratitude to LLNL for its support of this research. The authors acknowledge the work of Dr. Roy Mathew, Dr. Luigi Paduano, and Dr. Joseph Rard who were directly involved in the experimental measurements.

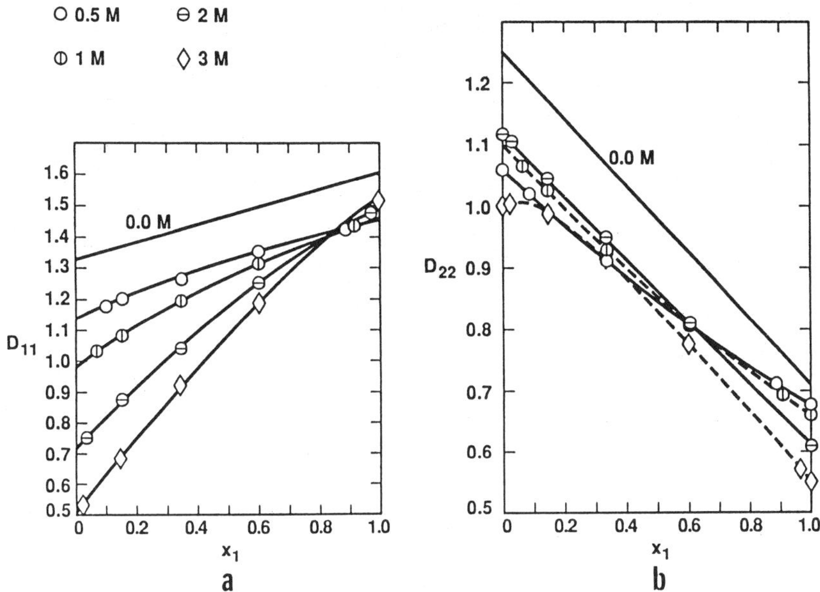

FIG. 1. (a) $10^9 D_{11}$ and (b) $10^9 D_{22}$ on the volume-fixed frame of reference versus equivalent fraction x_1 at various total molar concentrations. (Units: D_{ii}, $m^2 sec^{-1}$).

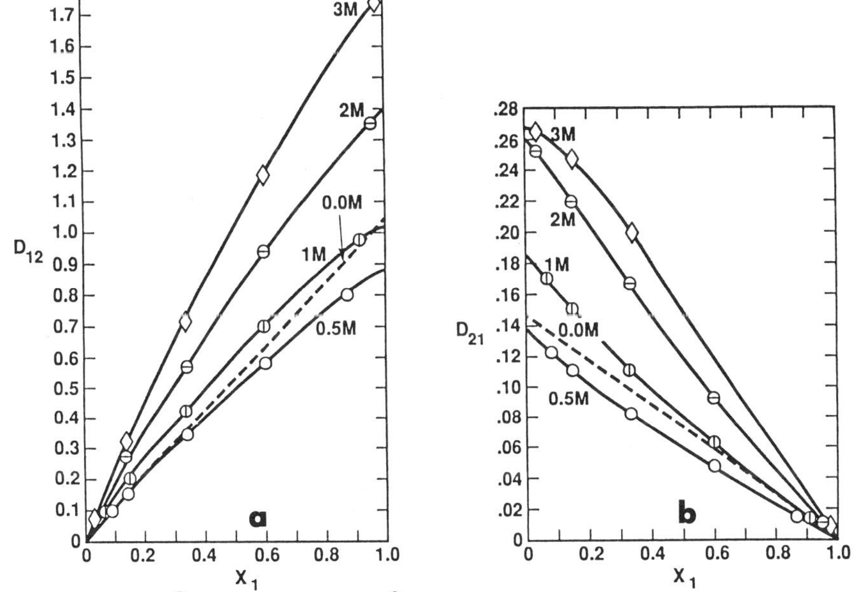

FIG. 2. (a) $10^9 D_{12}$ and (b) $10^9 D_{21}$ on the volume-fixed frame of reference versus equivalent fraction x_1 at various total molar concentrations. (Units: D_{ij}, $m^2 sec^{-1}$).

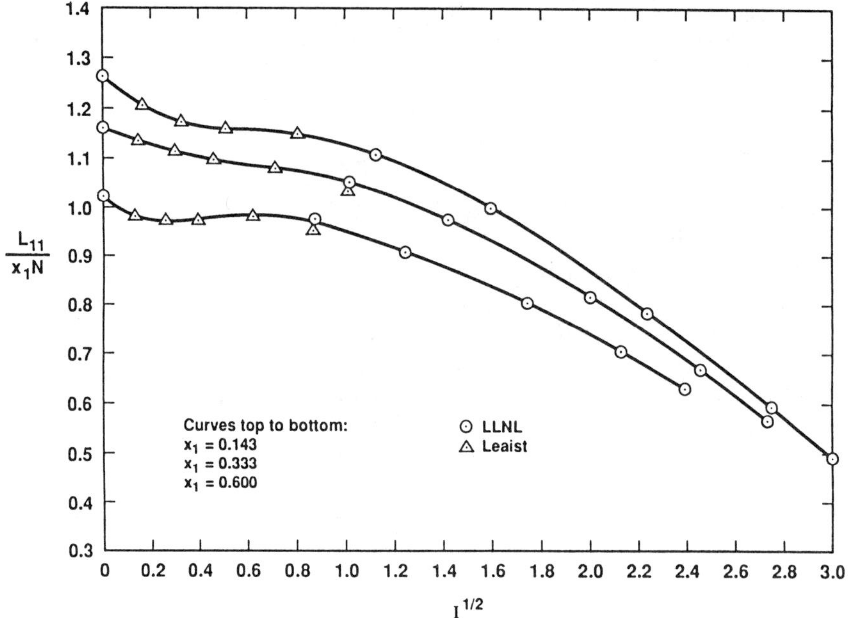

FIG. 3. 10^6RT $L_{11}/(x_1N)$ on the solvent-fixed frame of reference versus $I^{\frac{1}{2}}$. (Units: RT L_{11}, mol $m^{-1}sec^{-1}$; N, equiv dm^{-3}; I, mol dm^{-3}). The label L_{11} in the figure is an abbreviation for 10^6RT L_{11}.

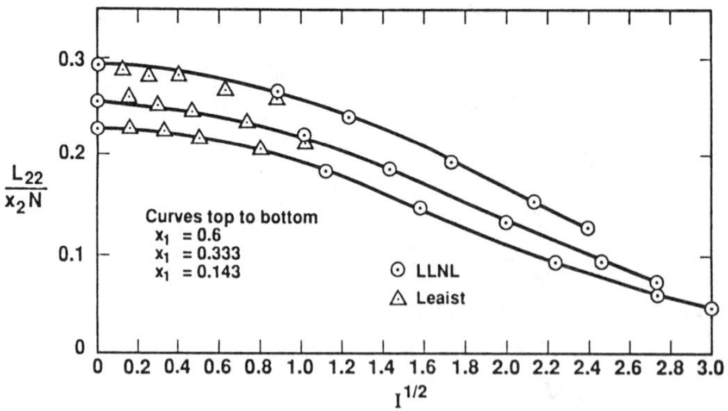

FIG. 4. 10^6RT $L_{22}/(x_2N)$ on the solvent-fixed frame of reference versus $I^{\frac{1}{2}}$. (Units: RT L22 mol $m^{-1}sec^{-1}$; N, equiv dm^{-3}; I, mol dm^{-3}). The label L_{22} in the figure is an abbreviation for 10^6RT L_{22}.

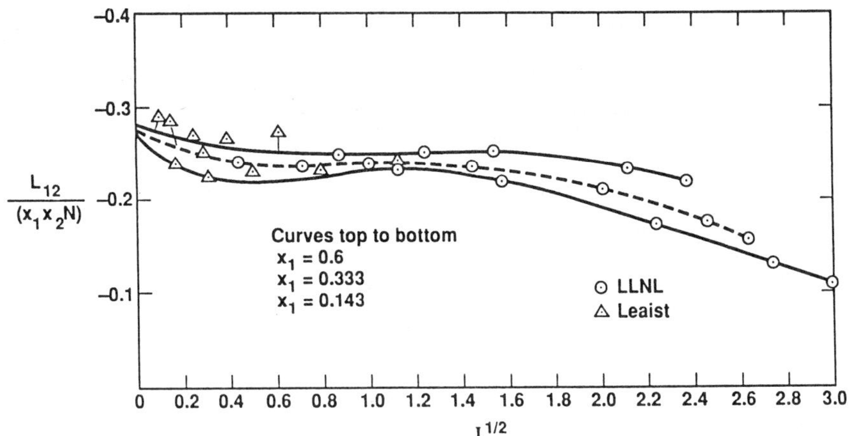

FIG. 5. 10^6RT $L_{12}/(x_1x_2N)$ on the solvent-fixed frame of reference versus $I^{½}$. (Units: RT L_{12} mol $m^{-1}sec^{-1}$; N, equiv dm^{-3}; I, mol dm^{-3}). The label L_{12} in the figure is an abbreviation for 10^6RT L_{12}.

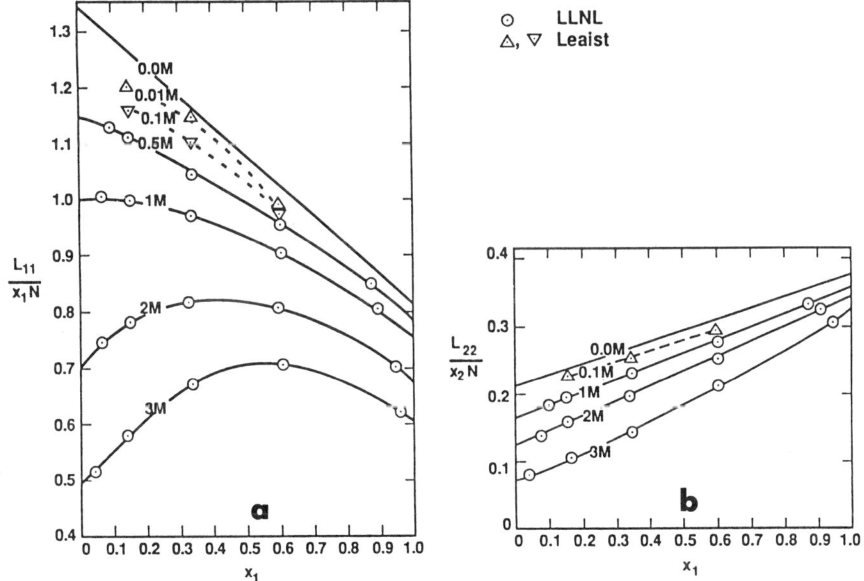

FIG. 6. (a) 10^6RT $L_{11}/(x_1N)$ and (b) 10^6RT $L_{22}/(x_2N)$ on the solvent-fixed frame of reference versus x_1 at various total molar concentrations. (Units: RT L_{ii} mol $m^{-1}sec^{-1}$; N, equiv dm^{-3}). The label L_{ii} in the figure is an abbreviation for 10^6RT L_{ii}.

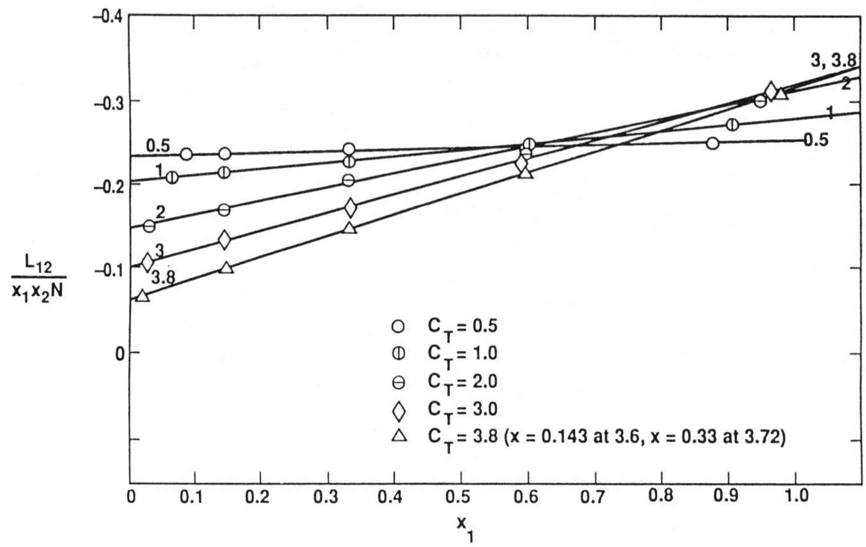

FIG. 7. 10^6RT $L_{12}/(x_1x_2N)$ on the solvent-fixed frame of reference versus x_1 at various total molar concentrations. (Units: RT L_{12} mol $m^{-1}sec^{-1}$; N, equiv dm^{-3}). The label L_{12} in the figure is an abbreviation for 10^6RT L_{12}.

References

1. L. J. Gosting, H. Kim, M. A. Loewenstein, G. Reinfelds and A. Revzin, Rev. Sci. Instrum. 44, 1602 (1973).
2. J. G. Albright, R. Mathew, D. G. Miller and J. A. Rard, J. Phys. Chem. 93, 2176 (1989).
3. L. Paduano, R. Mathew, J. G. Albright, D. G. Miller and J. A. Rard, J. Phys. Chem. 93, 4366 (1989).
4. R. Mathew, L. Paduano, J. G. Albright, D. G. Miller and J. A. Rard, J. Phys. Chem. 93, 4370 (1989).
5. D. G. Leaist, Electrochim. Acta 33, 795 (1988).
6. R. Mills, A. J. Easteal and L. A. Woolf, J. Solution Chem. 16, 835 (1987).
7. H. Bianchi, H. R. Corti and R. Fernández-Prini, J. Solution Chem. 18, 485 (1989).
8. J. A. Rard and D. G. Miller, J. Chem. Eng. Data 32, 85 (1987).
9. D. G. Miller, J. Phys. Chem. 71, 616 (1967).
10. W. J. Hamer and J.-C. Wu, J. Phys. Chem. Ref. Data 1, 1047 (1972).
11. J. A. Rard and D. G. Miller, J. Chem. Eng. Data 26, 38 (1981).
12. R. A. Robinson and R. H. Stokes "Electrolyte Solutions", 2nd ed. (revised); Butterworths: London 1965.
13. D. G. Miller, J. A. Rard, L. B. Eppstein and J. G. Albright, J. Phys. Chem. 88, 5739 (1984).

Periodic structures in multicomponent diffusion couples

F.J.J. Van Loo*

Eindhoven University of Technology, Laboratory of Solid State Chemistry and Materials Science, P.O. Box 513, 5600 MBE, Eindhoven, The Netherlands

K. Osinski

Phillips Research Laboratories, Way, P.O. Box 80000, 5600 JA, Eindhoven, The Netherlands

I - Introduction

Some years ago we reported on periodic structures, found in solid-state diffusion couples Fe_3Si-Zn and Co_2Si-Zn(1-2). Since then we acquired more experimental data on these couples, and besides we found the same type of structures formed in another system, viz.SiC-Ni(3). It is, therefore, worth while to evaluate again the occurrence of these structures, the more so since in a well-known textbook (4) our previous results are mentioned in a chapter on the Liesegang phenomenon. We shall point out, however, that these structures are formed in a different way. In fig.1 a photograph is shown of an Fe_3Si-Zn couple, annealed at 395°C(1). The intermetallic phases δ-$FeZn_{10}$ and ζ-$FeZn_{13}$ are formed, which are the zinc-rich phases also occurring in binary Fe-Zn couples. However, at very regular distances rows of FeSi are found in these phases. It turns out that these bands are in fact two-phase mixtures consisting of precipitates of FeSi between which the δ and ζ crystals continue to grow. The distance between these bands is very regular, and the thickness of the total reaction layer as well as the number of bands increase parabolically with time. After short annealing times, only one two-phase band is visible, still adhering to the substrate.

The differences with a Liesegang mechanism are conspicuous :

a- The distances Δx_n between the bands x_{n+1} and x_n are nearly constant, whereas in a Liesegang phenomenon they increase with increasing value of n (towards the substrate side) according to the Jablczynski relation $\Delta x_n/x_n$ = constant (4).

b- Immediately after a band has been released from the substrate, a new band is formed (Fig.2). This means that no nucleation problems, due to the lack of a required supersaturation, are present which would have been typical for Liesegang bands.

In this paper we present new results, which point to reasons of a mechanical nature for an explanation of the observed phenomena.

---

*Presently on leave at CNRS-IMP, Université de Perpignan, avenue de Villeneuve, 66025 Perpignan Cedex, France

II - Experimental Results

A large number of diffusion couples have been made in various ways, depending on the system. Mostly, the polished couple halves are pressed in a clamp or directly by weights in a vacuum furnace as described earlier (1-3). For couples which were meant to deform by stresses developping during the diffusion annealing (especially Fe_3Si-Zn), zinc was evaporated on one side of the Fe_3Si-substrate, or applied by hot dipping (2). The necessary alloys for preparing the couple halves have been made by argon-arc melting.

II - 1 The Fe-Si-Zn system

In the framework of this paper we will discuss the results on couples between solid zinc and the alloys Fe_3Si and Fe_5Si_3.
Starting with Fe_3Si-Zn couples (figs. 1 and 2) we confirmed the parabolic growth of the total layer and the parabolic increase of the number of FeSi-bands (showing the constancy of the average distance of about 16μm between these bands, see figs. 3 and 4).
On closer examination some details attract attention. After short annealing times, in some couples only the ζ-phase appears to be formed without a visible δ-layer, a phenomenon which is also often encountered in pure Fe-Zn couples (5). In this case the two-phase bands FeSi + ζ are thinner and closer together. If after longer annealing times the δ-phase is formed between ζ and the substrate, the FeSi + δ bands have their normal thickness and mutual distance as shown in fig. 5.
Since both the ζ and the δ layers grow during further annealing, inevitably FeSi-bands which originally were situated in the δ layer will be overtaken by the growing ζ-layer. Since during the transition from δ-$FeZn_{10}$ to ζ-$FeZn_{13}$ zinc is taken up, the distance between the bands increases by more than 10% as can be seen on close examination of figs. 1 and 5.

If the Fe_3Si-substrate is rather thin (150μm) and coated on only one side with a thin layer of zinc, the FeSi-bands grow thicker, adhere longer to the substrate and, therefore, their mutual distance is larger as seen from fig .6. The layer growth is generally less than in couples with thick Fe_3Si-substrates, but the retarding influence of the thicker FeSi-band on the total growth is not very large as can be seen in fig.7.
A specific characteristic of thin Fe_3Si-Zn couples is the possibility of deformation by mechanical stresses. We indeed found in most cases a small curvature in the couple, but this was not reproducible enough to draw conclusions from, the more so as also thermal mismatch stresses develop during cooling of the sample.
At higher temperatures (up to 450°C) liquid zinc is in contact with Fe_3Si. The periodic structure in the diffusion layer remains the same, and even the band distance remains the same despite the much faster layer growth.
Using Fe_5Si_3 as a substrate, the band formation is quite comparable with Fe_3Si-Zn couples as can be seen from figs. 3, 4 and 8, although in the longer

annealed couple (166 hours at 395°C) the bands are remarkably thick and far apart, especially near the substrate. In the rather brittle Fe_5Si_3-substrate a lot of cracks are present, which can be observed in the growing layer by the sudden breaks in the rhythm of the band formation (fig. 8).
An interesting remark with respect to Fe_5Si_3 is the metastabile state in which it can be used. The compound is stable only above 825°C, but its decomposition into Fe_3Si + FeSi is so slow that it cannot be observed even after long annealing times at 395°C. However, because of inhomogeneities formed during melting Fe_3Si crystals could be found in the Fe_5Si_3-matrix. At places where Zn reacts with both Fe_3Si and Fe_5Si_3, it is clear that the band formation is continuous along both substrate crystals although it is sometimes slightly thinner opposite to Fe_3Si as shown in fig 9. Besides, the composition of the bands are different. Those facing Fe_5Si_3 contain more FeSi as is comprehensible considering the chemical reaction equation. At places where FeSi precipitates, present in Fe_5Si_3, come in contact with the growing δ layer this growth stops, as can be predicted from the existing equilibrium between δ and FeSi.

II.2 The Ni-Si-C system

A comparable periodic structure is found in reactions between Ni and (Ni, Fe) alloys with SiC (see fig.10). The reaction layer consists of Ni-or (Ni,Fe)-silicides with precipitation bands of graphite at regular distances (3,6). In couples of pure Fe-SiC, graphite is found as very small and random particles in an Fe_3Si-matrix without any trace of band formation.

In this respect the (Ni, Fe)-SiC couples are interesting, since there band formation is seen in the Ni-rich silicides formed at the SiC-side, whereas in the Fe-rich silicides at the substrate side random precipitation of carbon is found (see fig.10). Since the reaction front is at the SiC/layer interface (Ni and Fe are the only diffusing species) the bands, formed in the Ni-rich silicides, are during further layer growth overtaken by the Fe-rich silicides and then they desintegrate into small random particles.

II.3 The Co-Si-Zn system

In couples of the type Co_2Si-Zn a reaction layer is found, consisting of a γ_2-$CoZn_{13}$ matrix with porous bands of CoSi(2). The band width and distance as observed in the microscope is not constant as for the FeSi bands in Fe_3Si-Zn couples, but seem to differ for each Co_2Si-grain from which the reaction layer grows (fig.11). It turns out that the CoSi bands actually are formed at probably the same distances, but that they release from the substrate under an angle, different for each Co_2Si grain. The width and distance variations seen in fig.11 are, therefore, only seemingly and the result of oblique cutting of the various "cells". In cells with the thinnest and most close-packed bands with distances of about 4μm, these are parallel to the original interface.

III - Discussion

It is clear that large resemblances but also differences are found between the three systems discussed here. The main point of correspondence is the formation of bands in an analogous way with respect to the phase diagram and diffusion mechanism. The phase diagrams and diffusion paths are shown in fig.12. After short annealing times the sequences of the phases in the couples are : $Fe_3Si/FeSi/FeZn_x/Zn$, $SiC/C/SiNi_x/Ni$ and $Co_2Si/CoSi/CoZn_x/Zn$. These layer sequences are predictable from the phase diagrams according to a thermodynamic model proposed by us (7). In all cases the substrate elements are immobile. The attacking elements (Zn and Ni) diffuse through the reaction layer towards the substrate and react with one element leaving behind the other one in pure form (like C) or as an intermetallic compound (FeSi and CoSi). These are formed as a porous layer at the substrate interface.

The overall compositions of these porous two-phase bands (about $Fe_{32}Si_{28}Zn_{40}$ for Fe_3Si-Zn, $Fe_{42}Si_{40}Zn_{18}$ for Fe_5Si_3-Zn and $Co_{34}Si_{33}Zn_{33}$ for Co_2Si-Zn couples) show that the Si fraction is about the same as in the substrate. This reflects the immobility of Si in these couples. The same is true for C in SiC-Ni couples.

This initial layer sequence is not governed by the phase relations only. Obviously other factors are important, too. The relative mobilities of the elements in the various phases is certainly one of them (7). Also the original composition of the substrate alloy is very important (2). For instance, if an Fe-Si substrate with less than 25% Si is used for the reaction with zinc, the formation of a continuous band of FeSi is impossible because of the slow supply by diffusion of the required silicon atoms. Another requirement is the capability of sintering of the reaction product at the substrate interface to a coherent porous layer, as was for instance shown in a study on the Cu_2o/Ni system (8). Obviously, the presence of Fe is not favourable for sintering of carbon and gives rise to very small and isolated carbon precipitates in Fe-SiC couples.

Once present, the periodic release of the porous layer from the substrate is the most intriguing aspect. If this were governed by the phase relations only, the phenomenon should have been found in more systems. We only found an analogous band formation in Fe(Ge)/Zn couples, albeit less clear than in the case of Fe_3Si/Zn. In the resembling couples Cu(Si)/Zn, Ni(Si)/Zn and a number of others we did not find a periodic band structure although we did not try to optimize the substrate composition up to now.

We feel that mechanical stresses built up during the reactions is the most important requisite for the formation of the periodic structure. The dependence of the band periodicity on the substrate thickness can hardly be explained in another way. There are more indications pointing into this direction.

a- The somewhat wavy appearance of the FeSi-bands definitely gives the impression that at a number of places the band blisters away from the substrate as some oxide layers do from metals like Ti(9,10), thus releasing the stored elastic energy in the strained band. Instead of a continuous oxygen gas phase like in the Ti oxidation, we now have e.g. a continuous $FeZn_{10}$ phase through which Zn may diffuse. A thin substrate releases this stored energy partly elastically, thus giving rise to thicker bands further apart.

b- Sometimes when the FeSi-band releases from the substrate the rupture does not occur at the interface substrate/band but somewhere in the middle of the porous band.

c- The thinner bands at shorter distances in the ζ-phase in Fe_3Si-Zn couples might also be related to mechanical-chemical factors. As can be seen from the phase diagram in fig. 12a, ζ cannot be in equilibrium with the substrate material Fe_3Si. Therefore, between the (FeSi + ζ)-band and the Fe_3Si-substrate the δ-phase must be formed, which might lead to extra volume changes and, therefore, larger stresses which cause a faster release of the band from the substrate.

d- In couples of the type Fe_5Si_3/Zn we often observed that thicker bands are formed on small Fe_5Si_3 parts which are separated from the brittle bulk couple half by cracks. We believe that less stresses occur at the interface with these small parts. Also the thicker layers and larger band distances near the substrate for the longer annealed and heavily cracked Fe_5Si_3/Zn couple point into this direction.

At the moment, we are trying to find quantitatively consistent evidence for the role of the various parameters on the process of periodic structures in diffusion couples.

IV Conclusions

We have demonstrated the inadequacy of the Liesegang mechanism to explain the periodic structures found in the diffusion couples Fe_3Si/Zn, Fe_5Si_3/Zn, Co_2Si/Zn and SiC/Ni. We have presented evidence for a growth stress mechanism causing the periodic release of a porous reaction layer from the substrate interface in a solid-state reaction couple.

IV - References

1 - K. Osinski, A. W. Vriend, G.F. Bastin and F. J. J. Van Loo, Z. Metallkde 73, 258 (1982).
2 - K. Osinski, The Influence of Aluminium and Silicon on the Reaction between Iron and Zinc, PhD-thesis Eindhoven University of Technology, the Netherlands, 1983.
3 - R.C.J. Schiepers, F.J.J. Van Loo and G. de With, J. Am. Ceram. Soc. 71, C-284 (1988).
4 - J.S. Kirkaldy and D.J. Young, Diffusion in the condensed state, The Institute of Metals, London, 1987.
5 - M. Onishi, Y. Wakamatsu and H. Miura, Trans. Jap. Inst. Mat. 15, 331 (1974).

6 - R.C.J. Schiepers, J.A. Van Beek, E. de Giacomoni, B. Vialla, F.J.J. Van Loo, and G. de With, Scripta Metallurgica, in print.
7 - F.J.J. Van Loo, J.A. Van Beek, G.F. Bastin, and R. Metselaer, Diffusion in Solids, M.A.Dayananda and G.E. Murch, Eds., TMS, Warrendale, pp.231-2(1985).
8 - F.J.J. Van Loo, P.J.C Vosters, J.G.M Becht and R. Metselaer, Materials Science Forum, 29, 261 (1988).
9 - J. Stringer, Acta Metall. 8, 758 (1960).
10 - L. Lattaud, D. Ciosmak and G. Bertrand, Reactivity of Solids 1, 57 (1985).

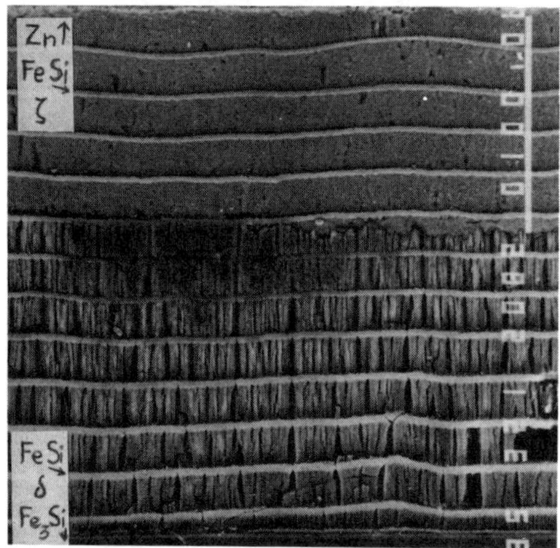

Fig. 1 - Fe_3Si-Zn couple, 24h, 395°C. The white bar indicates 100 μm.

Fig. 2 - Part of the reaction layer in an Fe-Zn couple, showing the immediate formation of a new band after the previous one has been released. Bar indicates 10 μm.

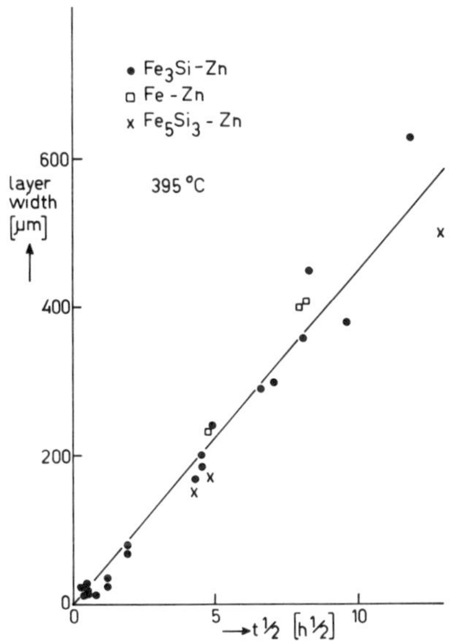

Fig. 3 - Plot of the layer width d (in µm) in various couples as a function oft $^{1/2}$ (in hours) at395°C; $d = 45t^{1/2}$.

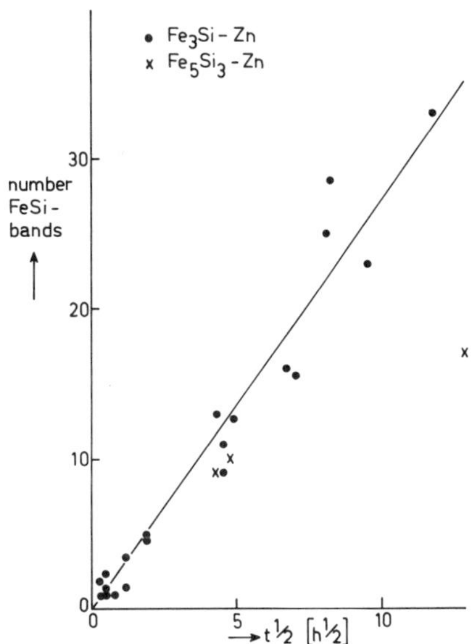

Fig. 4 - Plot of the number of FeSi-bands n as a function of t $^{1/2}$ (in hours) at 395°C; $n = 2.7t^{1/2}$.

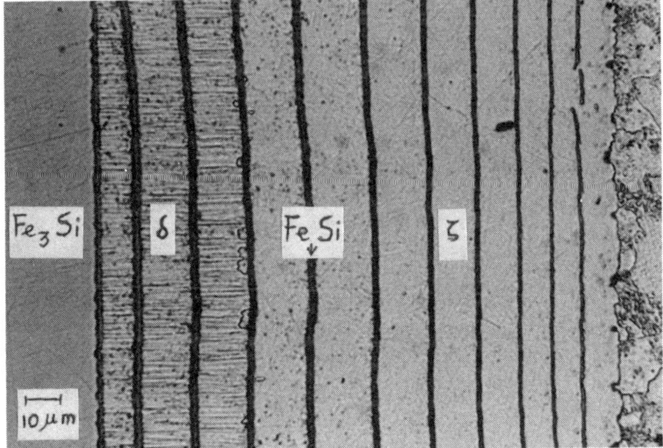

Fig. 5 - Couple Fe_3Si-Zn, 16h, 395°C. Note the differences in band width.

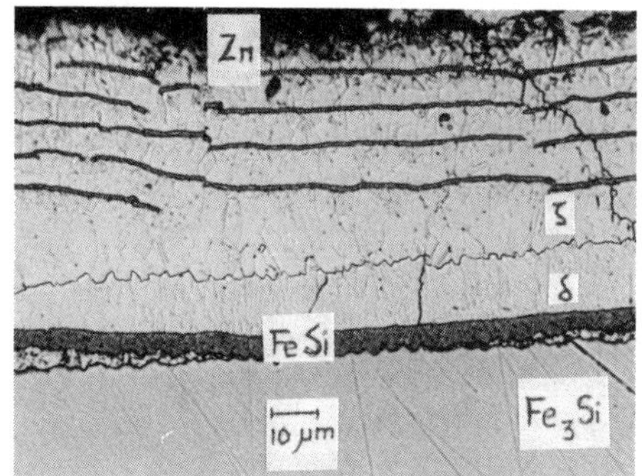

Fig. 6 - Thin substrate Fe_3Si-Zn couple, 24h, 395°C. Note the thick FeSi-band in the δ-phase.

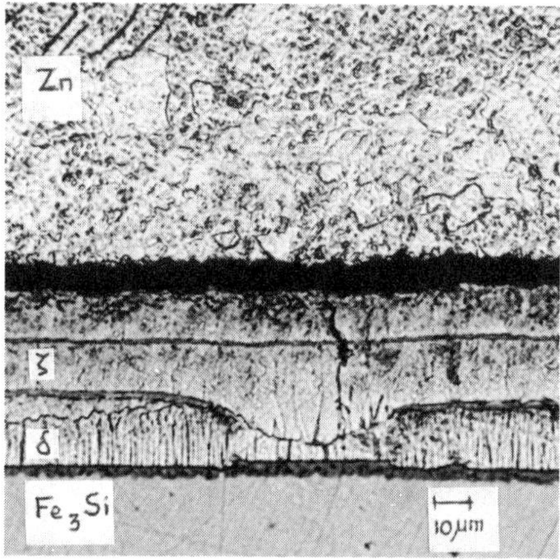

Fig. 7 - Part in an Fe_3Si-Zn couple, 16h, 395°C. Note differences in thickness of the band at the interface.

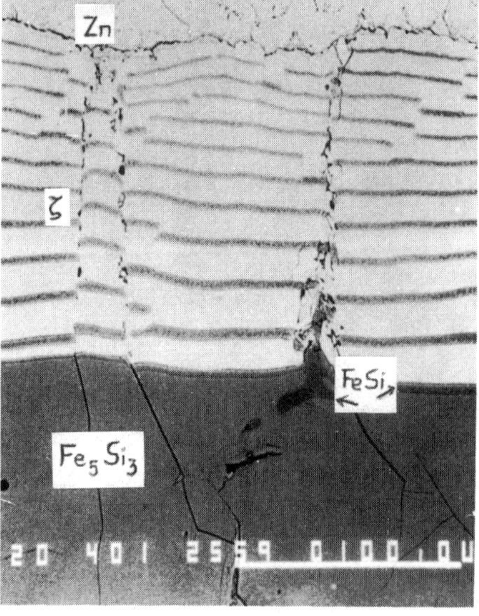

Fig. 8 - Fe_5Si_3-Zn couple, 18h, 395°C. Bar indicates 100 μm.

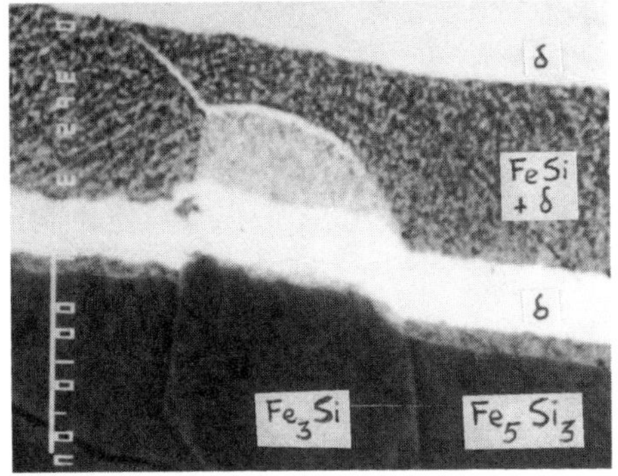

Fig. 9 - (Fe_5Si_3 + Fe_3Si)-Zn couple, 18h, 395°C. Note the difference in band formation into the Fe_5Si_3 and Fe_3Si substrates. Bar indicates 10 μm.

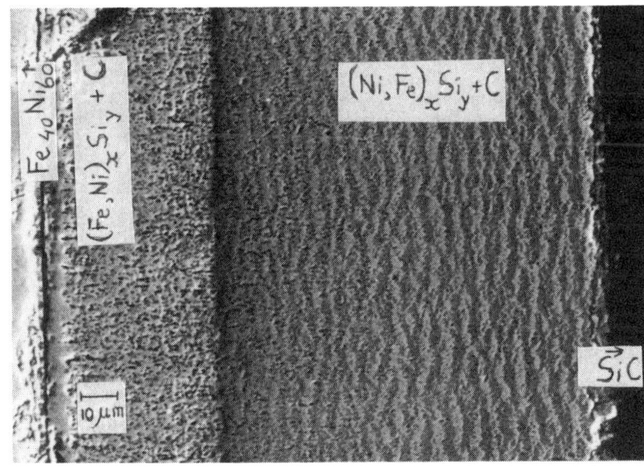

Fig. 10 - Fe_{40} Ni_{60}-SiC couple, 72h, 850°C. Note the absence of regular bands of graphite in the Fe-rich silicide.

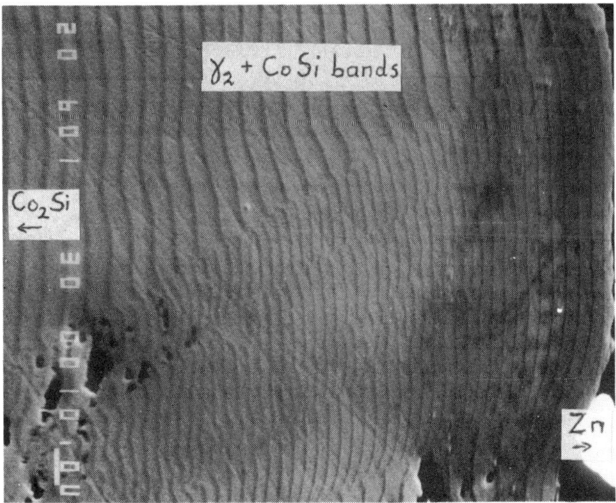

Fig. 11 - Co_2Si-Zn couple, 44h, 395°C. The white bar indicates 10 μm. Note the different orientation of CoSi-bands in the various "cells".

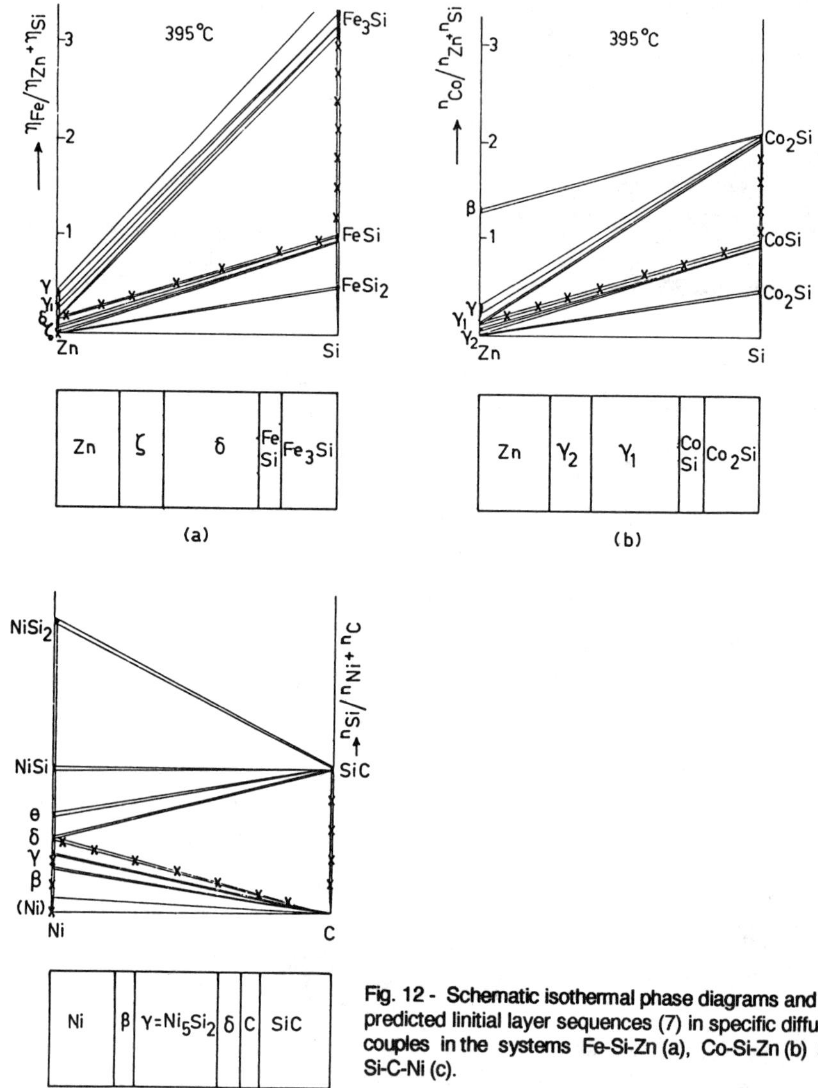

Fig. 12 - Schematic isothermal phase diagrams and the predicted linitial layer sequences (7) in specific diffusion couples in the systems Fe-Si-Zn (a), Co-Si-Zn (b) and Si-C-Ni (c).

Measuring the diffusivity of ternary, quaternary and higher order alloys

J.E. Morral, Yoon-Ho Son
Department of Metallurgy and Institute of Materials Science, University of Connecticut, U-136, Storrs, Connecticut 06269-3136, U.S.A.

M.S. Thompson
United Technologies Research Center, East Hartford, Connecticut 06108, U.S.A.

Introduction

The foundations of multicomponent diffusion theory can be found in a series of papers published by Jack Kirkaldy and coworkers between 1957 and 1966 in the Canadian Journal of Physics (1-6). The development of this field over the last thirty years coupled with the ability of modern computers to model complex diffusion controlled reactions has created a need for diffusion data on ternary, quaternary and higher order alloys.

There are two basic approaches that have been applied to measuring diffusivities from multicomponent diffusion couples. One has been to make couples containing large concentration differences and to analyze the data with a "variable diffusivity" analysis. Another approach has been to make couples containing small concentration differences and to analyze the data with a "constant diffusivity" analysis. These approaches and their advantages and disadvantages have been reviewed recently (7,8). In the present work three methods of analyzing couples with small concentration differences will be reviewed and compared.

Couples with large concentration differences will not be considered here because they have limited value for studying multicomponent systems. The problem is that multiple diffusion couples must be prepared which have crossing diffusion paths. The diffusivity is obtained at the composition of the crossing point, as explained by Kirkaldy, Lane and Mason (4). However, it can be seen from the geometry of diffusion paths, that they will normally cross only for ternary systems. In quaternary or higher order systems the diffusion paths, which are lines, can not be designed to cross except under special conditions (e.g. when the diffusivity is constant). Fig. 1 gives diffusion paths for ternary and quaternary systems in order to illustrate the problem.

Couples with small concentration differences are made in order to limit variations of the diffusivity across the diffusion zone. With sufficiently small variations, it can be assumed that the diffusivity is a constant. Under these conditions, error function solutions (i.e. "constant diffusivity" solutions) to the diffusion equation (9) predict that all diffusion paths will pass through the "average composition" (i.e. the interface composition) of the couple. Therefore, diffusion couples with the same average

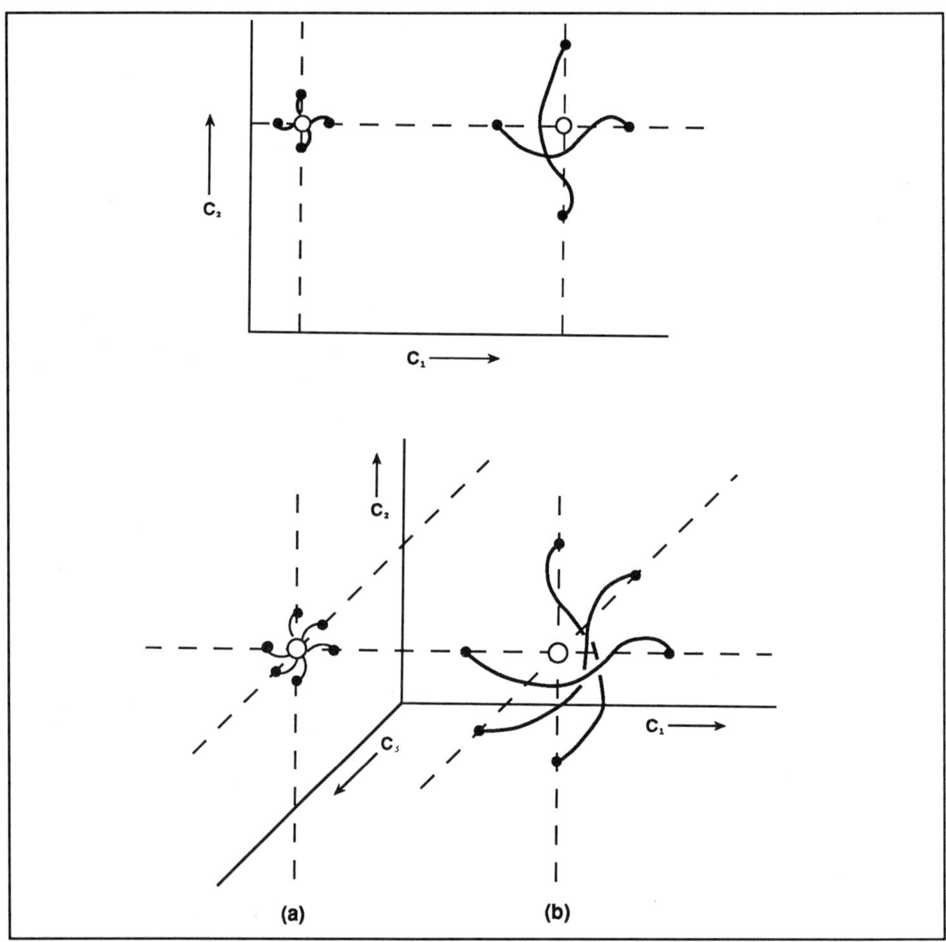

Fig. 1: Multicomponent diffusion paths for (a) constant diffusivity and (b) variable diffusivity diffusion couples. The average composition of each couple is given by open circles, o, while the initial couple alloys are given by closed circles, •.

composition will have diffusion paths that all cross at the average composition as is illustrated in Fig. 1. Other advantages associated with couples having small concentration differences are that variations in partial molar volumes, Kirkendall porosity, and sample distortion are all reduced (7).

Analysis of Couples with Small Concentration Differences

Three analyses will be given that can be used to analyze diffusion couples with small concentration differences. First is a multicomponent form of the Boltzmann-Matano analysis (7). It will be termed the BZMA analysis and it yields the diffusivity matrix, [D], directly. It is a general analysis that can be used when [D] varies, however, it requires constant diffusivity behavior in order to obtain crossing diffusion paths.

The other two analyses assume that [D] is constant in order to derive the analysis equations. One is by Krishtal, Mokrov, Akimov and Zakharov (10). It will be termed the KMAZ analysis and it yields the "inverse square root diffusivity" matrix, [p], which is related to [D] by the equation:

$$[D] = [p]^{-1}[p]^{-1} \tag{1}$$

The other constant [D] analysis will be termed the SQRD analysis and it yields the "square root diffusivity" matrix, [r], which is related to [D] by the equation (11-13):

$$[D] = [r]\,[r] \tag{2}$$

All three analyses give [D] for the average composition of the couple. In order to obtain [D] for other compositions it is necessary to prepare and analyze other diffusion couples.

Three measurements made on concentration profiles provide the data needed to calculate [D] by the three analyses. One measurement is of the amount of each solute, S_i, that has crossed the initial interface during the diffusion anneal. It is given by the expression:

$$S_i(t) = \int_0^\infty (C_i(x,t) - C_i^R)\,dx \tag{3}$$

Two other measurable quantities are the concentration gradient of each component at the original couple interface:

$$\nabla C_i = \partial C_i/\partial x]_{x=0} \tag{4}$$

and the difference in solute concentration between alloys on the left and right side of the interface defined by:

$$\Delta C_i^o = C_i^R - C_i^L \tag{5}$$

These measurements are illustrated on Fig. 2. It should be noted that S_i can be positive or negative depending on whether there is a net flux of solute entering or leaving the right side of the couple, respectively.

The basis of the BZMA analysis is a well known equation (7), which can be written in the terms defined above as:

$$S_i = -2t\sum_{1}^{n-1} D_{ij}\nabla C_j \tag{6}$$

As a general rule, (n-1) diffusion couples are required to obtain all elements of [D] from this equation. For example, for a ternary system (n=3) two couples are needed and these provide data for the following two equations written in terms of the amount of solute 1 which has crossed the initial interface:

$$S_1' = -2t(D_{11}\nabla C_1' + D_{12}\nabla C_2') \tag{7}$$

$$S_1'' = -2t(D_{11}\nabla C_1'' + D_{12}\nabla C_2'') \tag{8}$$

in which (') refers to data from one couple and (") refers to data from the other. Equations (6) and (7) can be solved simultaneously for D_{11} and D_{12}. Similar equations written for solute 2 can be solved to obtain D_{21} and D_{22}. Accordingly, for an n component system, (n-1) different diffusion couples with the same average composition provide (n-1) versions of equation 5 for each row of [D]. These can be solved simultaneously in order to obtain all elements of each row. Repeating the procedure for each row leads to the calculation of [D].

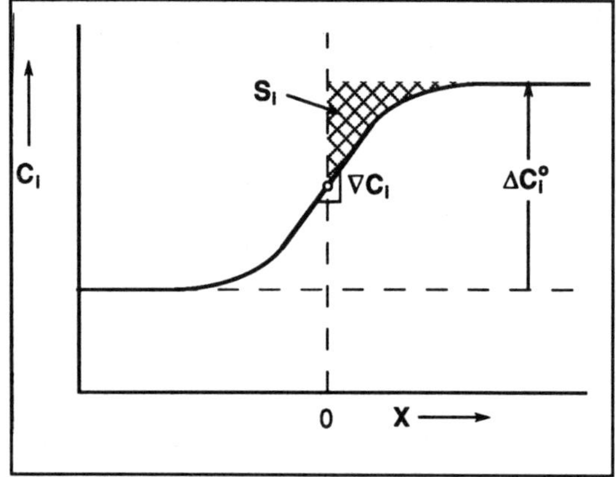

Fig. 2: Three measurements on a concentration profile that can be used to calculate diffusivities with "constant diffusivity" analysis methods.

The basis of the KMAK analysis is their equation 11 in reference (10) which relates concentration gradients at the initial interface to initial concentration differences in the diffusion couple.

In the present notation it is given by the following expression:

$$\nabla C_i = (2\sqrt{\pi t})^{-1} \sum_{1}^{n-1} p_{ij} \Delta C_j^o \tag{9}$$

The basis of the SQRD analysis is an equation which relates the solute crossing the initial interface with the initial concentration differences (12,13) and is given by:

$$S_i = -\sqrt{t/\pi}\sum_{1}^{n-1} r_{ij}\Delta C_j^o \qquad (10)$$

Equation 9 uses one set of data from the diffusion couples to solve for elements of [p] while equation 10 uses another set of data to solve for elements of [r].

The [p] and [r] matrices can be converted to [D] by equations 1 and 2. The result is three [D] matrices produced by the three analysis methods. The differences between the elements of these matrices provide a measure of the confidence limits.

Experimental Design and Procedure

There is an unlimited number of alloys that can be used in the (n-1) diffusion couples required to determine [D]. However, a systematic procedure can be followed that tends to distribute the diffusion paths uniformly when plotted on concentration coordinates as shown in Fig. 1. The procedure is to prepare couples like those designed by Darken (14) to study diffusion in the Fe-C-Si system. In his experiment the concentration of one solute (and the solvent) was different in the diffusion couple alloys, while the concentration of the other solute was the same.

Extending Darken's procedure to multicomponent alloys, the (n-1) diffusion couples are prepared so that each one has an initial concentration difference in just one of the (n-1) solutes present. All couples have the same average composition so that their diffusion paths will cross.

To insure uniform temperature the (n-1) couples can be clamped together before heat treatment. After a diffusion anneal, the assembly can be sectioned and placed in a single mount for microprobe analysis.

Results

Microprobe data of concentration versus distance is needed for each solute in each of the (n-1) couples. Therefore, a total of $(n-1)^2$ concentration profiles are required. For example in recent experiments at 1100°C on alloys having an average composition of Ni-9.7at%Cr-7.7at%Al, it was necessary to make two diffusion couples and measure four concentration profiles (15). The profiles were fit with simple functions which were then used to obtain values of S_i and ∇C_i. Values of ΔC°_i were obtained from unreacted parts of the couple. Substituting these results into equations 5-7 and then equations 1-2 yields the D_{ij}'s given in Table I.
Included in the table is the [D] reported by Nesbitt and Heckel (16) for the same composition. Their results were obtained by applying the Boltzmann-Matano analysis to a series of diffusion couples containing large concentration differences and then fitting the various D_{ij} values to a parabolic equation in concentration. The table lists the D_{ij} values calculated from the equation for Ni-9.7%Cr-7.7%Cr alloys.

TABLE I
Comparison of [D] Measured by Different Methods
for a Ni-9.7at%Cr-7.7at%Al alloy at 1100°C

	D_{AlAl}	D_{AlCr}	D_{CrAl}	D_{CrCr}
BZMA	23.7	8.1	7.4	11.5 x 10^{-11}cm^2/s
KMAZ	28.7	10.1	5.5	10.0
SQRD	22.0	7.6	7.8	12.6
NESH	23.0	7.3	6.3	9.4

BZMA Boltzmann-Matano Analysis
KMAZ Krishtal, Mokrov, Akimov and Zakharov Analysis
SQRD Square Root Diffusivity Analysis using couples with small concentration differences

NESH Nesbitt, Heckel results (16) using couples with large concentration differences

Discussion

The elements of [D] in Table I which were obtained by the "constant diffusivity" methods can be seen to vary from average values and the values obtained by Nesbitt and Heckel by about 10%. Two reasons can be given for the variation. One is the uncertainty in measurements of S_i, ∇C_i and ΔC°_i which was on the order of 10%, too. The other reason is that elements of [D] are, in principle and in fact, not constant with respect to composition. For example, Nesbitt and Heckel's equations predict that some of the D_{ij}'s varied by more than 10% over the concentration range of 4% which was present in the diffusion couples.

However, a technique used in analyzing the data tends to minimize the effect of variations in diffusivity. The technique involves combining concentration profiles from the left and right side of the couple. The average of these profiles is representative of a constant diffusivity system with average composition.

As an example, Fig. 3 illustrates the behavior of a binary system. In this example from Crank (17), D varies linearly with concentration by a factor of seven over the concentration range of the diffusion couple. The initial interface (x=0) is taken at the average concentration of the couple. The concentrations on the left are superimposed on those on the right by the following transformation:

$$C(x,t)-C^{o}-abs|C(x,t)-C^{L}| \tag{11}$$

for x>0 by:

$$C(x,t)-C^{o}=abs|C(x,t)-C^{R}| \tag{12}$$

Then $(C(x,t)-C^{\circ})/|C^{R}-C^{L}|)$ is plotted versus $|x|$. The profile for a "constant diffusivity" system (the diffusivity used is the average of the initial diffusivities of the variable system) is plotted on the figure as well. It can be seen that the constant diffusivity curve approximates the average of the left and right side curves.

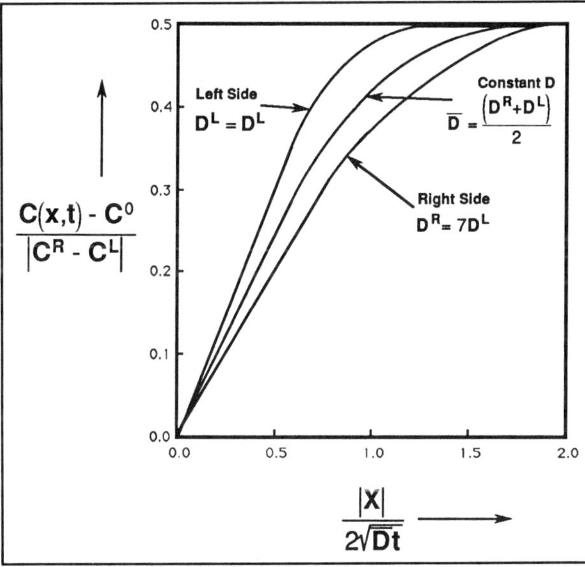

Fig. 3: Concentration profiles from the left and right side of a binary diffusion couple with "variable diffusivity" are combined and compared with a profile for a couple with "constant difusivity". The D's for each couple are given on the diagram.

The example infers that variations in diffusivity tend to broaden the data of the concentration profiles when the left and right side data are plotted together. However an average profile through the band can be analyzed in order to obtain the diffusivity of an alloy with average composition. Although, the example is for a binary system, a similar result would be expected for higher order systems in the limit of small initial differences in the concentrations.

Further evidence that meaningful diffusivities can be measured by constant diffusivity analysis methods has been obtained on both Ni-Cr-Al (15) and Ni-Cr-Al-Mo (18) systems. In these studies it was shown that measured diffusivities could be used to predict concentration profiles of Cr and Al to $\pm$.1 at%.

Conclusions

It is necessary to apply constant diffusivity analysis methods in order to measure diffusivities of quaternary and higher order systems, because variable diffusivity methods are limited by the requirement of crossing diffusion paths. The errors in the constant diffusivity analysis can be minimized by limiting the concentration differences in the couple and by averaging data from both sides of the diffusion couple.

Acknowledgements

The authors are grateful for support of this work by the National Science Foundation under grant number DMR-8711899.

References

1. J.S. Kirkaldy, Can.J.Phys. 35, 435 (1957); 36, 947 (1958); 37, 30 (1959).
2. J.S. Kirkaldy and G.R. Purdy, Can.J.Phys. 40, 208 (1962).
3. J.S. Kirkaldy, D. Weichert and Zia-Ul-Haq, Can.J.Phys. 41, 2166 (1963).
4. J.S. Kirkaldy, J.E. Lane and G.R. Mason, Can.J.Phys. 41, 2174 (1963).
5. J.E. Lane and J.S. Kirkaldy, Can.J.Phys. 42, 1643 (1964).
6. J.S. Kirkaldy and J.E. Lane, Can.J.Phys. 44, 2059 (1966)
7. J.S. Kirkaldy and D.J. Young, Diffusion in the Condensed State, p. 175, The Institute of Metals, London (1987).
8. J.E. Morral and M.S. Thompson, Diffusion Analysis and Applications, ed. A.D. Romig, Jr. and M.A. Dayananda, p. 35, TMS, Warrendale, PA (1989).
9. H. Fujita and L.J. Gosting, J.Am.Chem.Soc. 78, 1099 (1956).
10. M.A. Krishtal, A.P. Mokrov, V.K. Akimov and P.N. Zakharov, Fiz. metal. metalloved, 35, 1234 (1973).
11. J.E. Morral, Scripta Met. 18 (1984) 1251.
12. M.S. Thompson and J.E. Morral, Acta metall. 34 (1986) 339.
13. M.S. Thompson and J.E. Morral, Acta metall. 34 (1986) 2201.
14. L.S. Darken, Trans. AIME, 180, 430 (1949).
15. M.S. Thompson, J.E. Morral, A.D. Romig, Jr., Metall.Trans.A (accepted for publication, 4/90).
16. J.A. Nesbitt and R.W. Heckel, Metall.Trans.18A, 2061 (1987).
17. J. Crank, The Mathematics of Diffusion, 2nd ed., p. 386, Oxford Univ. Press, NY (1975).
18. M.K. Stalker, University of Connecticut, M.S. Thesis (1989).

A methodology for obtaining diffusion coefficients in a three-phase ternary couple: GaAs/nickel

C.-H. Jan, D. Swenson and Y.A. Chang
Department of Materials Science and Engineering, University of Wisconsin, Madison, Wisconsin 53706, U.S.A.

ABSTRACT

A methodology was presented for determining the interdiffusion coefficients of phases formed in a ternary diffusion couple using the measured growth rates and the concentration profiles across the couple. Using this methodology the interdiffusion coefficients of the T-Ni_3GaAs phase were obtained from several GaAs/Nickel couples. Assuming the cross-intrinsic diffusion coefficients to be negligible, relationships between intrinsic diffusion coefficients and interdiffusion coefficients were derived, and values were obtained for the three intrinsic diffusion coefficients of the T-phase. The intrinsic diffusivity for nickel was the largest and that for arsenic the smallest. These data were rationalized in terms of the structure of the T-phase.

1.0 Introduction

Two types of composite materials systems have emerged in recent years: layered materials for applications in the electronics industry and high-temperature composite materials for applications primarily in the aerospace industry. Both types of materials are multi-component systems and their ultimate performance depends on the stabilities at the interfaces of the layered structures and the structural composites. In this paper, we will focus on interdiffusional phenomena at the interfaces, where Professor J. S. Kirkaldy has made monumental contributions (1-5). Since intermediate phases often occur in binary, ternary and high-order systems, a knowledge of the formation and dissolution of these phases during the fabrication, subsequent annealing and eventual service treatment of these composites is essential. A solution to the growth of a two-phase binary system was first proposed by Wagner as given by Yost (6). This solution was then extended to three-phase binary systems by Castleman (7) and Kidson (8). Heckel and co-workers (9) and Metin, Ihal and Romig (10) among others have applied this solution to the growth of multi-phase binary systems.

A general solution for the layer growth of a ternary system was provided by Kirkaldy (1-3). This solution was applied by Kirkaldy and Brown (4) to study the growth of several two-phase copper-zinc-tin diffusion couples and was recently used by Nesbitt and Heckel (10) to study several two-phase nickel-chromium-aluminum diffusion couples. It has not been applied to the growth of multi-phase ternary diffusion couples. It is worth noting that Rapp, Ezis and Yurek (12) did calculate the kinetics of the displacement reaction $Fe + Cu_2O = 2Cu + FeO$. For electronic materials consisting of metal/semiconductor and composite materials consisting of metal/ceramic systems, the phases formed do not normally have large ranges of homogeneity. Determination of the interdiffusion coefficients via the Matano-Boltzmann analysis (14,15) is less feasible since accurate compositional variations within the single-phase regions are intrinsically difficult to measure experimentally.

Accordingly, a methodology is formulated in the present study to derive interdiffusion coefficients from the rates of layer growth of a semi-infinite ternary bulk diffusion couple and the concentration profile across the couple. The GaAs/Nickel couple is chosen as an example to illustrate the approach used. A detailed account of this study is given elsewhere (13); only the methodology used and the results of the analysis are presented here.

2.0 Ternary Diffusion Theory - Growth of a Three-Phase Ternary Diffusion Couple

Figure 1 shows the concentration profiles for the growth of a β-phase from a ternary α/γ diffusion couple and the corresponding ternary isothermal section at constant pressure. The nomenclatures used in the present study are given in Table I. The width of the β-phase layer as a function of time is (1-5,13)

Table I
Nomenclatures Used

Symbol	Meaning
C_1, C_2, C_3	Concentrations of components 1, 2, 3; $C_1+C_2+C_3=1$
C_i	Concentrations of component i with i = 1 or 2
D_1, D_2, D_3	Intrinsic diffusion coefficients of a ternary phase, neglecting the cross-intrinsic diffusion coefficients.
$\tilde{D}_{11}, \tilde{D}_{12}, \tilde{D}_{21}, \tilde{D}_{22}$	Interdiffusion coefficients of a ternary phase.
$\tilde{D}_{i1}, \tilde{D}_{i2}$	Interdiffusion coefficient of a ternary phase with i=1 or 2
$\tilde{J}_1, \tilde{J}_2, \tilde{J}_3$	Diffusive fluxes of components 1, 2, 3
$\tilde{J}_i$	Diffusive flux of component i with i=1 or 2
$\tilde{J}_i^{\beta,\alpha}$	The superscript β,α refers to the flux of component i in the β-phase at the α/β interface
$K_i^{\alpha,\beta}$	$\left[\frac{dC_i}{d\lambda}\right]^{\alpha,\beta} = \sqrt{t}\left[\frac{\partial C_i}{\partial x}\right]^{\alpha,\beta}$ with i = 1 or 2
$K_i^{\beta,\alpha}$	$\left[\frac{dC_i}{d\lambda}\right]^{\beta,\alpha} = \sqrt{t}\left[\frac{\partial C_i}{\partial x}\right]^{\beta,\alpha}$ with i = 1 or 2
P	Pressure
t	Time
T	Absolute temperature in K, °C = K - 273
w^β	Thickness of a β-phase growing from a α/γ couple = $\xi^{\beta,\gamma}-\xi^{\alpha,\beta}$
W^β	A parameter relating w^β and $\sqrt{t}$ = $[Z^{\beta,\gamma}-Z^{\alpha,\beta}]$
x	Distance coordinate
$\xi^{\alpha,\beta}$	Position of the α/β interface = $Z^{\alpha,\beta}\sqrt{t}$
$\xi^{\beta,\gamma}$	Position of the β/γ interface = $Z^{\beta,\gamma}\sqrt{t}$
$Z^{\alpha,\beta}$	A parameter relating $\xi^{\alpha,\beta}$ and $\sqrt{t}$ $= \frac{2[-\tilde{D}^{\alpha}_{i1}K_1^{\alpha,\beta} - \tilde{D}^{\alpha}_{i2}K_2^{\alpha,\beta} + \tilde{D}^{\beta}_{i1}K_1^{\beta,\alpha} + \tilde{D}^{\beta}_{i2}K_2^{\beta,\alpha}]}{[C_i^{\alpha,\beta} - C_i^{\beta,\alpha}]}$
$Z^{\beta,\gamma}$	A parameter relating $\xi^{\beta,\gamma}$ and $\sqrt{t}$ $= \frac{2[-\tilde{D}^{\beta}_{i1}K_1^{\beta,\gamma} - \tilde{D}^{\beta}_{i2}K_2^{\beta,\gamma} + \tilde{D}^{\gamma}_{i1}K_1^{\gamma,\beta} + \tilde{D}^{\gamma}_{i2}K_2^{\gamma,\beta}]}{[C_i^{\beta,\gamma} - C_i^{\gamma,\beta}]}$

$$w^{\beta} = [\xi^{\beta,\gamma} - \xi^{\alpha,\beta}] = Z^{\beta,\gamma}\sqrt{t} - Z^{\alpha,\beta}\sqrt{t} = W^{\beta}\sqrt{t} \qquad [1]$$

The above results can easily be adapted to the growth of an n-phase ternary system (1-5,13). It is evident from Eq. [1] that the growth of a phase β in a semi-infinite ternary diffusion couple depends on the interdiffusion coefficients of β and those of α and γ as well as the concentrations at the two interfaces and the ranges of homogeneity for these phases. It has often been found empirically that the growth of a phase from a diffusion couple is parabolic with time and that the temperature dependence of the parabolic rate constant follows Arrhenius behavior. However, there is no clear physical interpretation for the parabolic rate constants and activation energies obtained since they depend on the properties not only of the phase involved but also those of its neighboring phases.

Having the values of $\tilde{D}_{i1}$'s and $\tilde{D}_{i2}$'s for all the pertinent phases with i = 1,2, the compositions at the various interfaces and the ranges of homogeneity from accurate phase diagram data, Eq. [1] allows us to calculate the growth of phases in a semi-infinite ternary diffusion couple. However, as has been mentioned, for many systems of current technological interest such as Gallium-Metal-Arsenic for electronic materials or for multi-component structural materials, values of $\tilde{D}_{i1}$'s and $\tilde{D}_{i2}$'s are not known. As stated in the introduction, we wish to obtain the interdiffusion coefficients using Eq. [1], the growth rates of intermediate phases and the concentration profiles obtained experimentally from semi-infinite ternary GaAs/Metal diffusion couples. The methodology for obtaining these values for the system GaAs/Nickel will be presented below as an illustrative example.

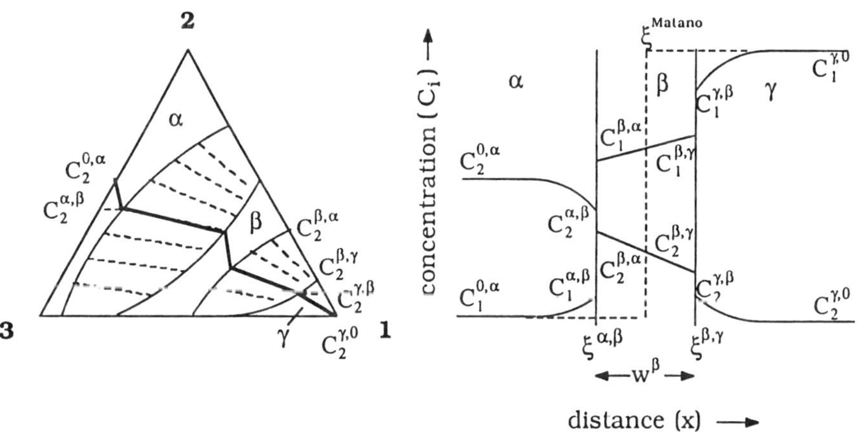

Fig. 1 The relationship between the diffusion zone interface concentrations for a three-phase ternary diffusion couple at constant temperature and pressure and the isothermal phase diagram for this ternary system.

3.0 Experimental Results and Methodology for Data Analysis: GaAs/Nickel; Formation of a Single Phase from the Couple

3.1 Methodology for Obtaining Interdiffusion Coefficients

When GaAs and nickel are brought into contact, a ternary phase with the nominal composition Ni_3GaAs, designated as T, is formed (16). A recent study of the phase equilibria of Gallium-Nickel-Arsenic at 600° C and the structure of the T-phase by Zheng, Lin, Swenson, Hsieh and Chang (17) showed T to be a stable phase with an appreciable range of homogeneity. It is neither in equilibrium with GaAs nor nickel. However, extensive bulk diffusion studies performed by us (13) in addition to the preliminary bulk diffusion data reported in Ref. 16 demonstrate that T is kinetically stable at 600° C for many days. In essence, metastable equilibrium is maintained between GaAs and T as well as between T and nickel. Moreover, the range of homogeneity of T is larger under this condition than that corresponding to equilibrium conditions (17), in accordance with thermodynamic arguments. Prior to the dissolution of T to other phases with longer annealing times, we can measure the growth of T as a function of time at different temperatures as well as the concentration profiles across the couples. The results obtained at 500, 600 and 700° C are presented in Fig. 2. In all of these diffusion couples, only the T-phase formed between the end phases, GaAs and nickel. At 600° C, a couple annealed for 11 days (264 hrs.) showed the formation of $Ni_{11}As_8$, Ni_5Ga_3, and Ni_3Ga between T and nickel; the T-phase also shrank. Longer annealing times may cause the complete disappearance of T. The growth of T in terms of its thickness, x, as a function of $t^{1/2}$ is given in Fig. 2. Within the scatter of the data, x varies linearly with $t^{1/2}$ for all three temperatures, indicating diffusion-controlled kinetics. The parabolic growth constants (i.e. slopes of the x-vs-$t^{1/2}$ curves) as discussed in Section 2 would normally depend on the interdiffusion coefficients of T and those of GaAs and nickel. In the following we will outline the methodology for obtaining the values of the interdiffusion coefficients of T from its growth rates and the concentration profiles across the diffusion couples.

As discussed in Section 2.0 and presented in Fig. 1, the growth of a three-phase ternary diffusion couple involves two interfaces. At each of these interfaces, there are two independent mass balance equations, yielding a total of four independent equations. In addition, we have one more equation, i.e. Eq. [1] for the growth of a particular phase β. However, for such a system we have a total of 12 interdiffusion coefficients, $\tilde{D}$'s, (four $\tilde{D}$'s for each of the three phases), and two interface positions, i.e. $\xi^{GaAs,T}$ and $\xi^{T,Ni}$. In order to solve this problem, there is a need to reduce the total number of unknown parameters to be determined. An examination of the

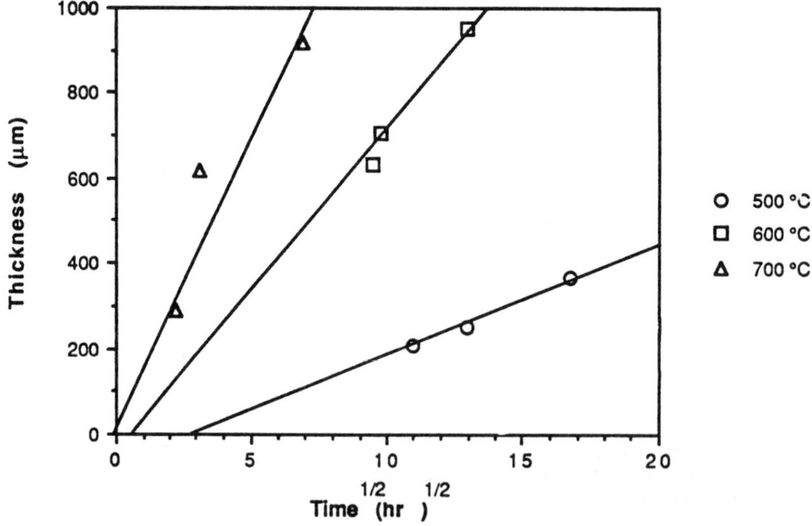

Fig. 2 The growth rates of T-Ni_3GaAs from GaAs/Ni bulk diffusion couples annealed at 500, 600 and 700° C.

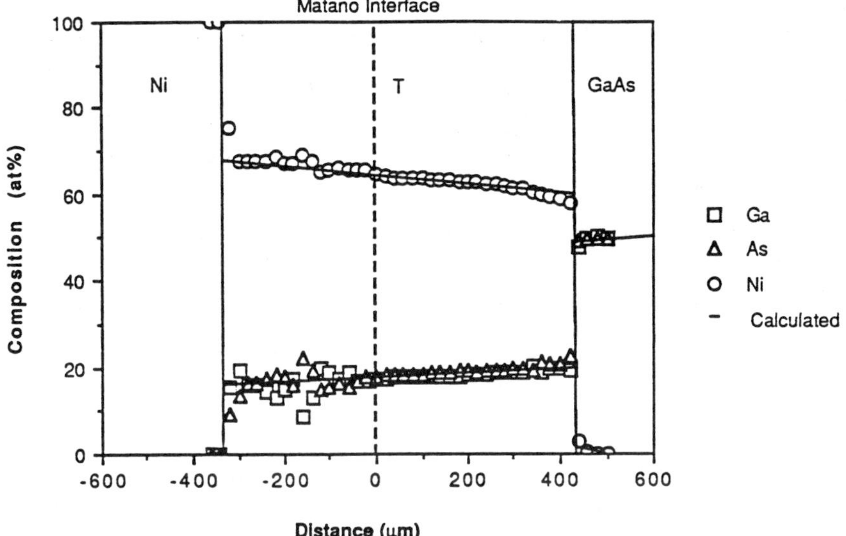

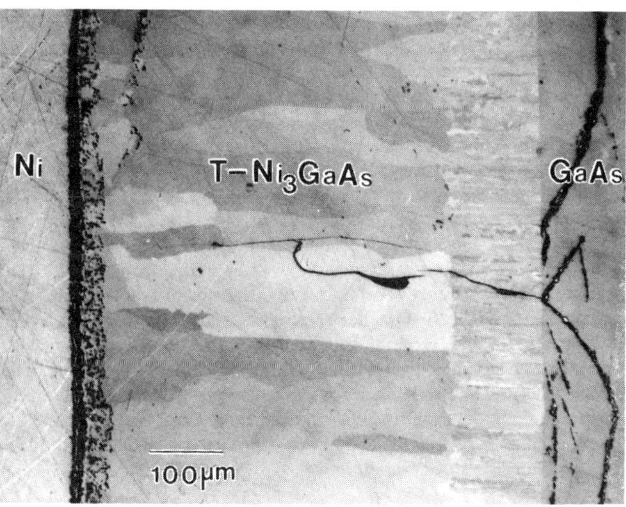

Fig. 3 a. GaAs/Ni bulk diffusion couple annealed at 600° C for 7 days (168 hrs.).

Microprobe linescan (the solid lines are calculated using Eq. [1] and the interdiffusion coefficients given in Table II).

b. Photomicrograph.

concentration profiles of all the GaAs/Nickel couples investigated (13) shows that the diffusive fluxes in the GaAs and nickel phases are much smaller than those in the T-phase. Otherwise, there would be penetration of the component elements in these two end phases. Accordingly, determination of the $\tilde{D}$'s in the GaAs and nickel phases is neglected. This leaves us only four unknowns: $\tilde{D}_{11}$, $\tilde{D}_{12}$, $\tilde{D}_{21}$, $\tilde{D}_{22}$. Figure 3a shows a typical concentration profile for a diffusion couple annealed at 600° C for 7 days (168 hrs.). Its corresponding photomicrograph is shown in Fig. 3b. Figure 3a shows negligible ternary penetration in the GaAs and nickel phases, justifying the assumptions made.

A review of the problem at hand shows that we have a total of five independent equations, i.e. the four interface mass balance equations and one for the growth of the T-phase. However, we have a total of six unknowns, which are $\tilde{D}_{11}$, $\tilde{D}_{12}$, $\tilde{D}_{21}$, $\tilde{D}_{22}$, $\xi^{GaAs,T}$ and $\xi^{T,Ni}$. It is therefore necessary to make a further assumption which will reduce the total number of unknowns to five or fewer. As a first approximation, we assume that the movement of one atomic species will not be influenced by the motion of another; i.e., there are no interdiffusional cross-terms. Therefore, $\tilde{D}_{12}$ and $\tilde{D}_{21}$ are assumed to be zero, and the remaining interdiffusion coefficients may be obtained by an optimization process, using the parabolic rate constants shown in Fig. 2.

Values of $\tilde{D}_{11}$ and $\tilde{D}_{22}$ with 1 being nickel and 2 arsenic are listed in Table II and presented in Fig. 4 in terms of an Arrhenius type of plot. Within the uncertainties of the data, values of $\tilde{D}_{11}$ and $\tilde{D}_{22}$ obtained decrease linearly with 1000/T. These two values do not differ appreciably from each other, and approach each other at low temperatures. This implies that we can almost treat this problem as a pseudobinary diffusion couple, reducing the number of interdiffusion coefficients to one. The activation energies obtained for the two interdiffusion coefficients are also given in Table II. As shown in Fig. 3a, the calculated concentration profiles of nickel and arsenic are in agreement with the experimental data.

Using the assumption that interdiffusional cross-terms are negligible is perhaps the most common approach to solving higher-order diffusion problems. In the following section, we will demonstrate an alternative approach to solving such problems, based upon relationships between interdiffusion and intrinsic diffusion coefficients. It will be shown that the intrinsic diffusion coefficients may be obtained directly from the solutions to the diffusion equations, and the interdiffusion coefficients found indirectly by back-substitution of the intrinsic coefficients into the equations relating the two types of coefficients.

3.2.1 Methodology for Obtaining Intrinsic Diffusion Coefficients and Interdiffusion Coefficients

The methodology presented in the previous sections allows us to obtain interdiffusion coefficients from the growth rates of phases and the concentration profiles across semi-infinite ternary diffusion couples. Once we have the interdiffusion coefficients of the phases involved, we can predict their growth rates as a function of time at a specified temperature. This is extremely useful for predicting quantitatively the extent of interaction between phases in a layered structure for applications in electronic materials or composite materials for structural applications. Nevertheless, interdiffusion coefficients, along with the theory of diffusion can only give us information concerning the phenomenological aspects of diffusion. In order to gain an understanding of the microscopic phenomena of diffusion, it is necessary to know the intrinsic diffusion coefficients. In the following, we will attempt to obtain values of the intrinsic diffusion coefficients using the same experimental data by making reasonable assumptions. This analysis will also allow us to obtain values for the interdiffusion coefficients in T-Ni_3GaAs, including the interdiffusional cross-terms.

In view of the fact that most of the phases in GaAs/Metal or fiber/metal systems are of a limited range of homogeneity, we may assume as a first approximation that all of the cross-intrinsic diffusion coefficients are negligible. In other words, three intrinsic diffusion coefficients, i.e. D_1, D_2 and D_3, are sufficient to describe the behavior of a ternary phase. Following the analysis of Darken (18) relating the interdiffusion coefficients to intrinsic diffusion coefficients for a binary system, the following relations for a ternary phase are obtained:

Table II

Interdiffusion Coefficients, $\widetilde{D}_{11}$ and $\widetilde{D}_{22}$ with 1 being Ni and 2 As

T, °C	$\widetilde{D}_{11}$	$\widetilde{D}_{22}$
500	$2.45(\pm 0.2)\times 10^{-9}$	$3.15(\pm 0.3)\times 10^{-9}$
600	$1.15(\pm 0.1)\times 10^{-8}$	$1.55(\pm 0.1)\times 10^{-8}$
700	$4.61(\pm 0.7)\times 10^{-8}$	$6.82(\pm 1.0)\times 10^{-8}$
Q(kJ/mol)	91.4±7	95.6±7

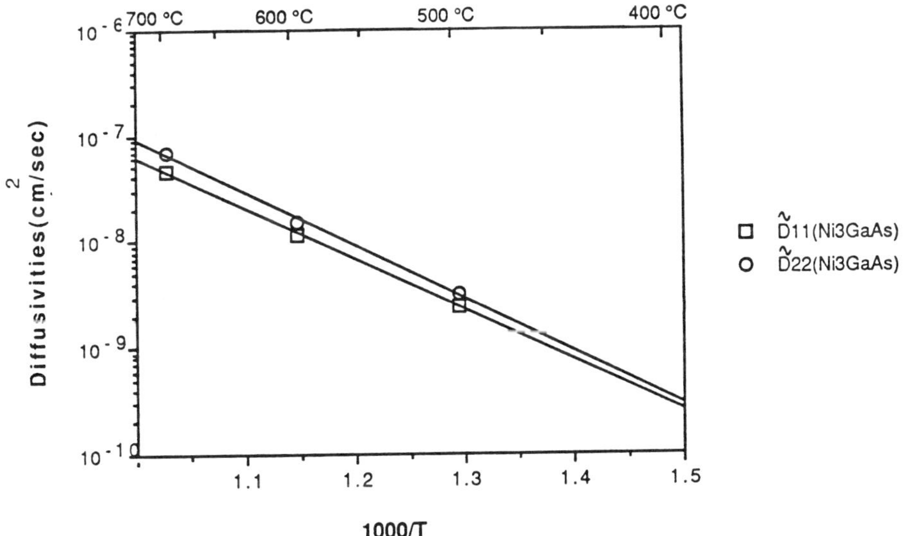

Fig. 4 **Arrhenius plot of $\widetilde{D}_{11}$ and $\widetilde{D}_{22}$ with 1 being Ni and 2 As; the activation energies are 91.4 and 95.6 kJ/mol respectively.**

$$\tilde{D}_{11} = D_1 + x_1 (D_3 - D_1) \quad [2]$$

$$\tilde{D}_{12} = x_1 (D_3 - D_2) \quad [3]$$

$$\tilde{D}_{21} = x_2 (D_3 - D_1) \quad [4]$$

$$\tilde{D}_{22} = D_2 + x_2 (D_3 - D_2) \quad [5]$$

These equations are consistent with those reported in the literature for ideal ternary solid solutions in which the "vacancy wind" effect is minimized (19,20).

In Section 3.1, values for $\tilde{D}_{11}$ and $\tilde{D}_{22}$ were obtained by using experimentally determined thicknesses and composition profiles of the T-Ni_3GaAs phase in GaAs/Nickel diffusion couples. It was necessary to set the interdiffusional cross-terms equal to zero in order to reduce the number of unknowns to five. However, Eqs. [2]-[5] automatically reduce four of these unknowns, the interdiffusion coefficients, to three unknowns, i.e. the intrinsic diffusion coefficients (assuming that experimental values have been obtained for x_1 and x_2 in Eqs. [2]-[5]). Equations [2]-[5] may therefore be substituted for the interdiffusion coefficients, and the five equations may be solved for the five unknowns $\xi^{GaAs,T}$, $\xi^{T,Ni}$, D_1, D_2 and D_3, without any assumptions being made as to the values of the interdiffusional cross-terms. Furthermore, the values obtained for the intrinsic diffusion coefficients may be back-substituted into Eqs. [2]-[5] to yield values for $\tilde{D}_{11}$, $\tilde{D}_{22}$ and $\tilde{D}_{12}$, $\tilde{D}_{21}$.

Although this methodology can be used to obtain numerical values for the interdiffusional cross-terms, it should be noted that this analysis was performed under the assumption of negligible cross-intrinsic diffusion coefficients, which is reasonable for phases with limited ranges of homogeneity. However, the phase T-Ni_3GaAs does possess an appreciable range of homogeneity, and therefore this assumption may be somewhat simplistic. Furthermore, it should be remembered that in performing these calculations, the diffusive fluxes in the nickel and GaAs phases were assumed to approach zero, as was the case in Section 3.1.

The resulting values for the intrinsic diffusion coefficients and interdiffusion coefficients of T-Ni_3GaAs are given in Tables III and IV and presented in Figs. 5 and 6, respectively. These results show that nickel is the fastest diffusing species, gallium the next fastest and arsenic the slowest. The intrinsic diffusion coefficient of nickel is two orders of magnitude larger than that of arsenic at 500° C. These data will be rationalized in terms of the structure of the T-phase.

3.2.2 Structural Interpretation of Intrinsic Diffusion in T-Ni_3GaAs

In the present study, large intrinsic diffusion coefficients and a small activation energy were found for the diffusion of nickel in the T-Ni_3GaAs phase. Such properties are characteristic of interstitial diffusion mechanisms. An examination of the crystal structure of the T-phase is therefore warranted to determine whether such a diffusion mechanism is reasonable for this system.

The crystal structure of T-Ni_3GaAs has been determined recently by Sands[21,22] and Zheng et al.[17]. The T phase possesses a $B8_{"1.5"}$ crystal structure, which is an intermediate of the hexagonal $B8_1$(NiAs) and $B8_2$(Ni_2In) structures. In this structure, as is shown in Figs. 7a and 7b, one of the two equivalent interstitial sites is randomly occupied by a nickel atom. (In the $B8_1$ structure, both interstitial sites are vacant; in the $B8_2$ structure, both are occupied).

Compounds possessing the B8 crystal structure are comprised of a transition metal in combination with a metalloid or Group B atom. They are metallic in nature, particularly when their axial ("c"/"a") ratios are much smaller than the ideal ratio for hexagonal close-packing, $(8/3)^{1/2} = 1.63$[23]. (For T-Ni_3GaAs, "c"/"a" = 1.30). Indeed, the bond lengths in T-Ni_3GaAs between nickel and gallium(arsenic) and nickel and nickel are consistent with Pauling's analysis for metallic bonding based upon fractional bond numbers[24]. Using Pauling's equations or simple geometric calculations, metallic radii of 0.125 nm and 0.131 nm may be assigned to nickel and

Table III

Intrinsic Diffusion Coefficients, D_1^T, D_2^T and D_3^T with 1 being Ni, 2 As and 3 Ga

T, °C	D_1^T	D_2^T	D_3^T
500	$8.72(\pm0.7)\times10^{-9}$	$1.37(\pm0.1)\times10^{-10}$	$3.46(\pm0.3)\times10^{-10}$
600	$3.65(\pm0.3)\times10^{-8}$	$2.86(\pm0.2)\times10^{-9}$	$6.70(\pm0.6)\times10^{-9}$
700	$1.17(\pm0.1)\times10^{-7}$	$2.58(\pm0.2)\times10^{-8}$	$7.32(\pm0.5)\times10^{-8}$
Q(kJ/mol)	81±6	163±11	167±11

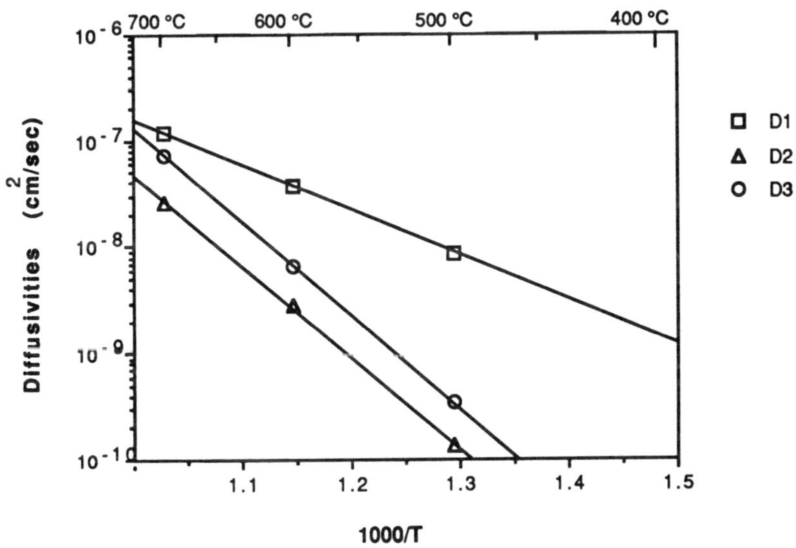

Fig. 5 Arrhenius plot of D_1, D_2 and D_3 with 1 being Ni, 2 As and 3 Ga; the activation energies are 81, 163 and 167 kJ/mol respectively.

Table IV

Interdiffusion Coefficients, $\tilde{D}_{11}$, $\tilde{D}_{12}$, $\tilde{D}_{21}$ and $\tilde{D}_{22}$ with 1 being Ni and 2 As (Determined from Intrinsic Diffusion Coefficients)

T, °C	$\tilde{D}_{11}$	$\tilde{D}_{12}$	$\tilde{D}_{21}$	$\tilde{D}_{22}$
500	$3.22(\pm0.3)\times10^{-9}$	$1.37(\pm0.1)\times10^{-10}$	$-1.44\pm0.1)\times10^{-9}$	$1.73(\pm0.4)\times10^{-10}$
600	$1.73(\pm0.3)\times10^{-8}$	$2.47(\pm0.2)\times10^{-9}$	$-5.33(\pm0.4)\times10^{-9}$	$3.54(\pm0.3)\times10^{-9}$
700	$9.01(\pm0.6)\times10^{-8}$	$2.91(\pm0.2)\times10^{-8}$	$-8.48(\pm0.8)\times10^{-9}$	$3.49(\pm0.3)\times10^{-8}$
Q(kJ/mol)	103±7	167±12	56±4	166±12

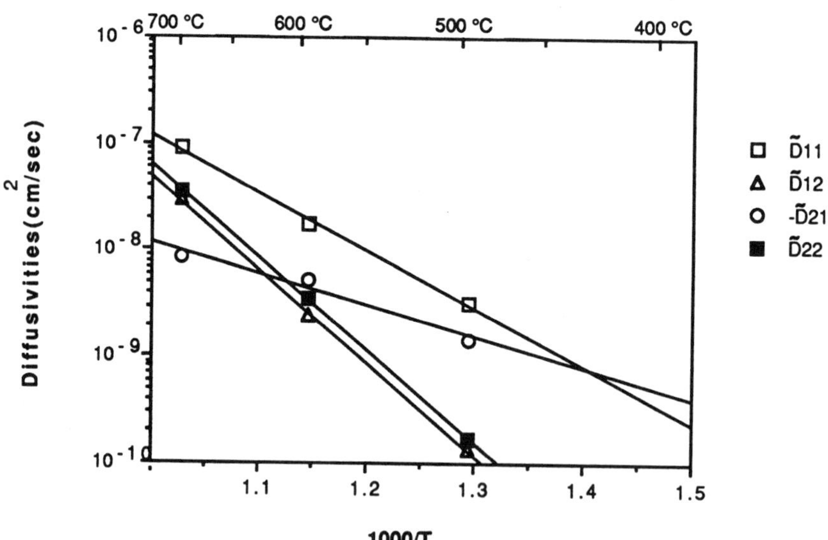

Fig. 6 Arrhenius plot of $\tilde{D}_{11}$, $\tilde{D}_{12}$, $\tilde{D}_{21}$ and $\tilde{D}_{22}$ with 1 being Ni and 2 As (determined by using intrinsic diffusivities); the activation energies are 103, 167, 56 and 166 kJ/mole, respectively.

gallium(arsenic), respectively. These radii may be used to calculate the radius of the unoccupied interstitial site in the T-Ni_3GaAs unit cell. The site is found to have a radius of 0.93 nm, a value which is 0.74 times the radius of nickel and 0.71 times the radius of gallium(arsenic) in this crystal structure.

These ratios may be compared to the ratios of interstitial radii to atomic radii in systems for which an interstitial diffusion mechanism has been established. In the well-known system (γ-iron)-carbon, for example, the radius of an octahedral interstitial site is 0.67 times that of a carbon atom [25]. Therefore, the assumption that diffusion in T-Ni_3GaAs is dominated by an interstitial mechanism seems very reasonable.

A specific interstitial diffusion mechanism has been proposed for compounds possessing B8 crystal structures [26]. In this two-step diffusion mechanism, an atom jumps from a lattice site (such as the origin of the unit cell (0,0,0)) to an interstitial site, (2/3,1/3,1/4) as shown in Fig. 8. It then jumps to one of six equivalent lattice sites: (0,0,0), (1,0,0), (1,1,0), (0,0,1/2), (1,0,1/2) or (1,1,1/2). The final position of the atom is referenced with respect to the origin of the unit cell. Although the atom diffuses the same total distance (namely, twice the distance from (0,0,0) to (2/3,1/3,1/4)) in each of the six jumps described above, an inspection of Fig. 8 shows that the net displacement of the atom with respect to the origin of the unit cell is not identical for each two-step jump. When the position vectors for each of the six jumps are projected on to the two orthogonal directions in the hexagonal unit cell (i.e. <100>, or the "a" direction and <001> or the "c" direction) it is found that overall the contributions to atomic displacement in the <100> and <001> directions are not equivalent, i.e. diffusion is anisotropic. Using the approximation $D_{<uvw>} = x^2/t$, where x is the jump distance in the <uvw> direction as measured from the origin of the unit cell, it may be shown that $D_{<100>}/D_{<001>} = 2"a"^2/"c"^2$ [26]. (In the case of T-Ni_3GaAs, $2"a"^2/"c"^2 = 1.18$).

The morphology of the T-phase in bulk GaAs/Nickel diffusion couples suggests an orientation relationship between T-Ni_3GaAs and GaAs (see Fig. 3b). In this couple, the T phase is characterized by fine, columnar grains which are parallel to the growth direction of this phase. It is likely that the anisotropy of diffusion in the T-phase contributes to this morphology, with directions in which diffusion is fast being parallel to the axes of the grains. Unfortunately, the orientation relation between T-Ni_3GaAs and GaAs has not been determined experimentally for bulk couples, and further study is therefore needed to confirm these speculations.

The preceding analysis of interstitial diffusion was formulated to rationalize the diffusion of transition metals in B8 phases. The diffusion of metalloid or Group B atoms was thought to be restricted to the metalloid sublattice [26]. Such diffusion would occur via intrinsic metalloid sublattice vacancies, and would therefore be slow compared to an interstitial mechanism. This would explain the diffusion behavior observed in the T-Ni_3GaAs phase, where the intrinsic diffusivity of nickel is greater than that of gallium or arsenic.

One might expect that at higher temperatures thermal vibrations would allow the gallium and arsenic atoms to escape temporarily the metalloid sublattice, i.e. some gallium and arsenic atoms would be able to occupy interstitial sites. This in turn would change the diffusion mechanisms of gallium and arsenic in this phase and would lead to intrinsic diffusivities of gallium and arsenic in T-Ni_3GaAs which would approach that of nickel. Such a relative increase in the intrinsic diffusivities of gallium and arsenic with respect to that of nickel was observed in the present study.

The fact that the intrinsic diffusivity of arsenic is roughly 2.5 times smaller than that of gallium over the range of temperature studied may be attributable to the electronegativity of arsenic. If the phase T-Ni_3GaAs possessed any ionic character, the arsenic atoms would act as anions, and would try to situate themselves in such a way as to maximize their distances from each other. Such an arrangement is realized in the B8"1.5" structure of T-Ni_3GaAs, where arsenic atoms have no other arsenic atoms as nearest neighbors or even as second-nearest or third-nearest neighbors. The nearest neighbor configurations for arsenic atoms in this phase are shown in Figs. 9. Figures 10 depict the atomic environment of an interstitial site in T-Ni_3GaAs. It may be seen that the configurations of nearest neighbors for arsenic atoms and interstitial sites are identical in terms of geometry, but have different types of atoms as first and second-nearest neighbors. As is shown in Figs. 10, if an arsenic atom were to occupy an interstitial site, it would have three arsenic atoms as nearest neighbors or two arsenic atoms as second-nearest neighbors, depending upon the position of the interstitial site, this configuration being

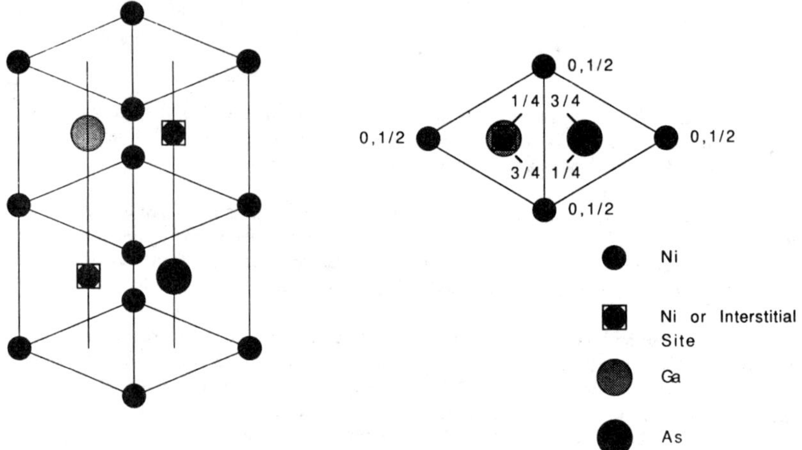

Fig. 7 a. Unit cell of the phase T-Ni_3GaAs (B8"$_{1.5}$" structure).
b. T-Ni_3GaAs unit cell as viewed along the "c" axis.

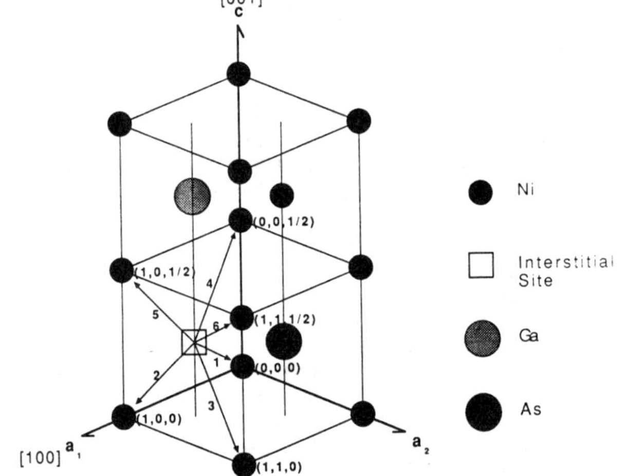

Fig. 8 Illustration of the two-step interstitial diffusion mechanism for nickel atoms in T-Ni_3GaAs. Upon jumping from (0,0,0) to an interstitial site at (2/3,1/3,1/4), an atom may jump to any one of sixe equivalent lattice sites (jumps labelled 1-6).

energetically unfavorable, due to the mutual repulsion of the arsenic atoms. Therefore, arsenic atoms might be somewhat more confined to the metalloid sublattice than gallium atoms, which are cationic in character and closer to nickel in electronegativity. The ability of gallium to occupy interstitial sites would translate into enhanced intrinsic diffusion of gallium with respect to arsenic. In most GaAs/metal systems, the diffusion of arsenic is found to be slower than that of metal or gallium, which lends credence to the ideas discussed above.

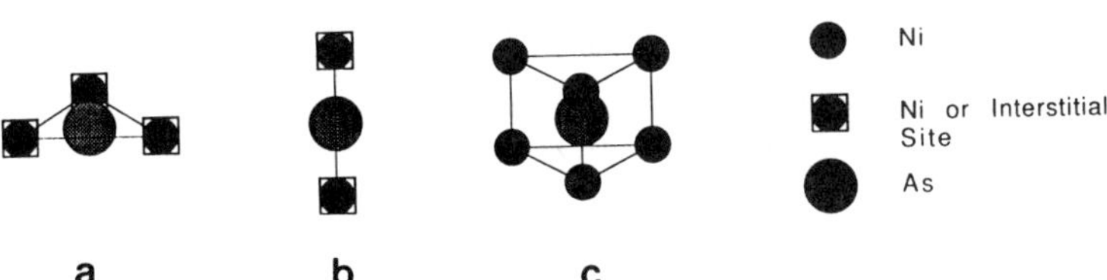

Fig. 9 Nearest neighbor configurations for arsenic in T-Ni_3GaAs:
a. One and one half nearest neighbors (nickel).
b. One second-nearest neighbor (nickel).
c. Six third-nearest neighbors (nickel).

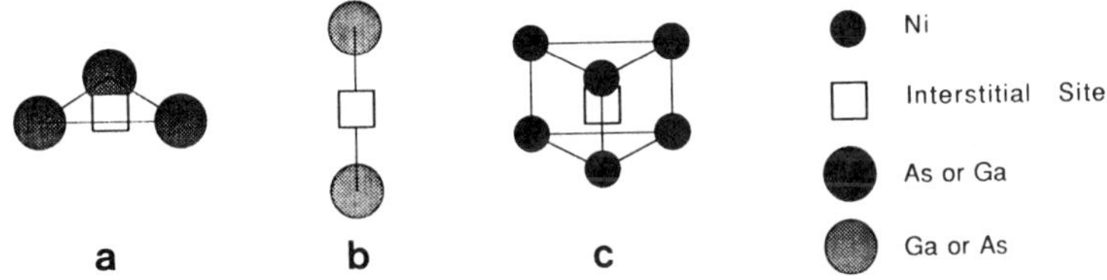

Fig. 10 Nearest neighbor configurations for an interstitial site in T-Ni_3GaAs;
a. Three nearest neighbors (arsenic (gallium)).
b. Two second-nearest neighbors (gallium (arsenic)).
c. Six third-nearest neighbors (nickel).

It is instructive to compare the intrinsic diffusion coefficients calculated for T-Ni_3GaAs with the experimental diffusion data of a similar phase. Such data is available for Ni_3Sn_2, which crystallizes in the B8"$_{1.5}$" structure at high temperatures and has a "c"/"a" ratio similar to that of Ni_3GaAs [27]. Intrinsic diffusion data for both nickel and tin were calculated from values of D_o and Q obtained by Schmidt [28] and are presented in Table V. Data were normalized to the congruent melting point of Ni_3Sn_2 (1264° C) [29] so as to correspond to the data calculated for T-Ni_3GaAs, which was estimated to melt at 900° C.

The intrinsic diffusivity of nickel in T-Ni_3GaAs is consistently larger than the diffusivity of nickel in Ni_3Sn_2. The ratio $D_{Ni}(T)/D_{Ni}(Ni_3Sn_2)$ varies from about 10^3 at $T/T_m = 0.66$ (500 and 740° C for T-Ni_3GaAs and Ni_3Sn_2, respectively) to 10^2 at $T/T_m = 0.83$ (700 and 1002° C for Ni_3GaAs and Ni_3Sn_2, respectively). The activation energy for the diffusion of nickel in T-Ni_3GaAs is also smaller than that of nickel in Ni_3Sn_2 by a factor of about 2.5.

There are some similarities in the intrinsic diffusion behavior of the two phases. For example, the ratios $D_{Ni}/D_{Ga(As)}$ in T-Ni_3GaAs and D_{Ni}/D_{Sn} in Ni_3Sn_2 decrease by an order of magnitude from $T/T_m = 0.66$ to $T/T_m = 0.83$, indicating that the intrinsic diffusivities of nickel and tin in Ni_3Sn_2 rapidly approach each other at high temperatures, as do the intrinsic diffusivities of nickel, gallium and arsenic in T-Ni_3GaAs.

The diffusion data given in Table V show that there is a reasonable agreement (one or two orders of magnitude) between calculated and experimentally determined intrinsic diffusivity data for compounds crystallizing in the B8"$_{1.5}$" structure, considering the simplistic treatment of intrinsic ternary diffusion used in calculating the diffusion coefficients and the likely differences in bond type and bond strength between nickel and tin and nickel and gallium (arsenic). It should also be noted that the temperature range over which the intrinsic diffusivities of nickel and tin were measured in Ni_3Sn_2 was not given by Schmidt.

Table V

Intrinsic Diffusion Coefficients, $D_{Ni}^{Ni_3Sn_2}$ and $D_{Sn}^{Ni_3Sn_2}$ (from Ref. 28)

T, °C	$D_{Ni}^{Ni_3Sn_2}$	$D_{Sn}^{Ni_3Sn_2}$
740	$8.62x10^{-12}$	$8.05x10^{-15}$
871	$1.54x10^{-10}$	$4.16x10^{-13}$
1002	$1.52x10^{-9}$	$9.53x10^{-12}$
Q(kJ/mol)	212	290

4.0 Conclusions

It has been demonstrated that the growth rates of intermediate phases and the experimentally determined concentration profiles across the phases may be used to determine their interdiffusion coefficients, which may in turn be utilized to calculate the growth of phases in a semi-infinite ternary diffusion couple as a function of time for a given temperature. Furthermore, this analysis may be used to find intrinsic diffusion coefficients in each phase, which may be helpful in developing atomistic interpretations of diffusion in these phases.

Specifically, the analysis outlined above was applied to the growth of T-Ni_3GaAs in semi-infinite, GaAs/Nickel diffusion couples to determine the interdiffusion coefficients for this phase. In performing the calculations, determination of the interdiffusion coefficients for nickel and GaAs was neglected, and the cross-term diffusivities for T-Ni_3GaAs ($\tilde{D}_{12}$ and $\tilde{D}_{21}$) were assumed to be vanishingly small. Further calculations of the growth of the T-Ni_3GaAs phase in bulk GaAs/Nickel diffusion couples using the calculated diffusivities were in good agreement with experimental data.

By assuming cross-intrinsic diffusional terms to be negligible, relationships between interdiffusion coefficients and intrinsic diffusion coefficients were derived, and the intrinsic diffusivities of nickel, gallium, and arsenic in T-Ni_3GaAs were determined. The intrinsic coefficients were then substituted into these relationships, yielding new values of interdiffusion coefficients, including cross-interdiffusional terms. The intrinsic diffusivity of arsenic was found to be two orders of magnitude smaller than that of nickel at 500° C. These intrinsic diffusivities were rationalized in terms of the crystal structure of the T-Ni_3GaAs phase (B8"$_{1.5}$"), which possesses large interstitial sites. An interstitial mechanism was postulated to explain the rapid diffusion of nickel within the phase, whereas electronegativity considerations were thought to explain the relative immobility of arsenic. This interstitial mechanism predicts anisotropic diffusion behavior, implying the possibility of an orientation relationship between T-Ni_3GaAs and GaAs in bulk GaAs/Nickel diffusion couples.

5.0 Acknowledgements

The authors wish to thank the Department of Energy for its financial support through Grant No. DE-F902-86ER-452754.

6.0 References

1. J. S. Kirkaldy, Can. J. Phys., 1958, 36, 917.
2. J. S. Kirkaldy, Can. J. Phys., 1958, 36, 899.
3. J. S. Kirkaldy, Can. J. Phys., 1958, 36, 907.
4. J. S. Kirkaldy and L. C. Brown, Can. Met. Q., 1963, 2, 89.
5. J. S. Kirkaldy and D. J. Young, Diffusion in the Condensed State, the Institute of Metals, London, 1987.
6. W. Jost, Diffusion in Solids, Liquids, and Gases, 2nd Ed., Academic Press, New York, 1960, p. 68.
7. L. S. Castleman, Nucl. Sci. Engin., 1958, 4, 209.
8. G. V. Kidson, J. Nucl. Mater., 1961, 3, 21.
9. A. J. Hickl and R. W. Heckel, Metall. Trans., 1975, 6A, 431.
10. E. Metin, O. T. Ihal and A. D. Romig, Jr., private communication from A. D. Romig, Jr., 1989.
11. J. A. Nesbitt and R. W. Heckel, Metall. Trans., 1987, 18A, 2087.
12. R. A. Rapp, A. Ezis and G. J. Yurek, Metall. Trans., 1973, 4, 1283.
13. C. H. Jan, D. Swenson, X.-Y. Zheng, J.-C. Lin and Y. A. Chang, Acta. Met., 1990, to be published.
14. L. Boltzmann, Ann. Physik, Leipzig, 1894, 53, 595.
15. C. Matano, Jpn. J. Phys., 1930, 8, 109.
16. J.-C. Lin, X.-Y. Zheng, K.-C. Hsieh and Y. A. Chang, in Epitaxy of Semiconductor Layered Structures, (Eds.: R. T. Tung, L. R. Dawson and R. L. Gunshor), MRS Symposium Proc., 1988, 102, 233.
17. X.-Y. Zheng, J.-C. Lin, D. Swenson, K.-C. Hsieh and Y. A. Chang, Mats. Sci. Engin., 1989, B5, 63.
18. L. S. Darken, Trans. AIME, 1949, 180, 430.
19. L. S. Castleman, Metall. Trans., 1983, 14A, 45.
20. J. R. Manning, Metall. Trans., 1970, 1, 499.
21. T. Sands, V. G. Keramidas, J. Washburn, and R. Gronsky, Appl. Phys. Lett., 1986, 48, 402.
22. T. Sands, V. G. Keramidas, A. J. Yu, K.-M. Yu, R. Gronsky and J. Washburn, J. Mater. Res., 1987, 2, 262.
23. A. Kjekhus and W. B. Pearson in Progress in Solid State Chemistry (Ed.: H. Reiss), v. 1, p. 83, 1964, Pergamon Press, Inc., Oxford, England.
24. L. Pauling, The Nature of the Chemical Bond, 1960, Cornell University Press, Ithaca, New York.
25. D. A. Porter and K. E. Easterling, Phase Transformations in Metals and Alloys, 1981, van Nostrand Reinhold (UK) Co., Ltd., Wokingham, England.
26. R. Hähnel, W. Miekeley and H. Wever, Phys. Stat. Sol. A, 1986, 97, 181.
27. M. Ellner, J. Less-Common Met., 1976, 48, 21.
28. H. Schmidt, Dissertation, TU, Berlin, 1989; H. Wever, J. Hüneke and G. Frohberg, Z. Metallkde., 1989, 80, 389; J. Hüneke, G. Frohberg and H. Wever, Defect and Diffusion Forum, 1989, 66-69, 483.
29. T. Heumann, Z. Metallkde., 1943, 35, 206.

Solidification of ternary peritectic alloys

M.G. Julian and G.R. Purdy
Department of Materials Science and Engineering, McMaster University, Hamilton, Ontario, Canada, L8S 4M1

Abstract

Solute partitioning during any phase transformation in a multi-component system is controlled by two factors: solute chemical interaction and solute diffusion kinetics for the phases present. A conceptual model has been developed which allows one to consider separately the influence of solute chemical interaction and diffusion kinetics on microsegregation. To better understand the role of solute diffusion kinetics on the growth of the peritectic phase during solidification, the solute diffusivities were assumed to be different by at least two orders of magnitude in the peritectic phase, i.e. Fe-C-X, where X is a substitutional solute. The thermodynamics of the system is described by the dilute solution and local equilibrium models; solute chemical interaction is described by Wagner interaction parameters. The description of the diffusional growth of the peritectic phase has bee simplified by assuming 1-D diffusion with complete mixing in the primary and liquid phases; off-diagonal diffusion coefficients are assumed to be negligible. Several methods are available for solving the non-isothermal diffusion problem, i.e. finite difference methods of Green's Function. In this work, the non-isothermal solidification-peritectic transformation process is modelled using a stepwise approach instantaneous temperature drops and isothermal diffusion for a specified length of time determined by the cooling rate versus time relationship.

Two limiting cases were considered: 1. the diffusion kinetics of the fast diffusing solute controls peritetic phase growth and the slow diffusing solute partitioning is described by the Scheil-Chipman equation. 2. the diffusion kinetics of the slow diffusing solute controls peritectic phase growth. Case one is shown to be self-consistent. The extremes in diffusivity that exist among the solute in the phases present in systems such as Fe-C-X suggest that the limiting case one predictions of phase composition and interface position would be a fair approximation to that for the real system. Within the context of this analysis, the system is self-moderating, the interactions between the fast and slow diffusing solute keeps the fast diffuser's peritectic concentration gradient negligibly small throughout the solidification/ transformation process. The slow diffuser's diffusion kinetics subtly influences the process through the choice of equilibrium interface concentrations.

3. Applications

Multicomponent diffusion interactions in gels: zeolites, polymers and slags

D. Okongwu
College of Engineering, Unviersity of Argiculture, Makurdi, Nigeria, MB 2378

J.S. Kirkaldy
Institute for Materials Research, McMaster University, Hamilton, Ontario, Canada, L8S 4M1

Abstract

Very large diffusion cross-interactions have been reported for impurities injected cocurrently into zeolites and polymers. This can be understood on the basis of Langmuir-type adsorption on the walls of the channels in these loosely packed structures. The theory of Karger and Bulow indicates that for a binary 1, 2 injection, the ratios of off-diagonal diffusion coefficients

$$\frac{D_{12}}{D_{11}} = \frac{X_1}{X_m - X_2} \; ; \; \frac{D_{21}}{D_{22}} = \frac{X_2}{X_m - X_1}$$

Hence if the channel site saturation concentration X_m is very small then the effective interaction parameter $1/(X_m - X_i)$ can be very large and D_{12}/D_{11} easily exceed unity. When the channels of open structures are wetted, a Kirkendall convective back-flow of the aqueous solution may be created which can affect the symmetry between the above relations to the extent of a sign change in one of the ratios. When $D_{11} >> D_{22}$ in a finite system reversion of the injected fast diffuser tends to occur. The relevance of these considerations to slag-metal reactions is discussed.

Introduction

Zeolites and polymers fall within the general materials class of gels, which for various reasons of chemical bonding are channelled or open structures as seen by diffusing species. Some metallurgical slags and membranes fall within the same class. One might expect some commonality of diffusion behaviour with solid state interstitial, grain boundary and phase boundary solutions in close-packed solids and this proves to be the case. This paper undertakes a synthesis along these lines, dealing among other things with the phenomenon of reversion in zeolites and slags. In particular we review the current understanding and some observations on zeolites and polymers, noting similarities and contrasts with other open structures. The distinctions are often associated with differences in characteristic boundary conditions rather than with matters of principle.

Adsorption of Binary Gases in the Channels of Zeolites

A typical zeolite structure is shown in Fig. 1 (1). This defines channels of the order of 2 nm cross section, of the same order as occurs in polymeric or biological membranes. Zeolites and intercalation compounds (the two dimensional analogue) are operated in the wet or dry condition, the latter as a dessicant involving gas adsorption and desorption. Polymeric membranes are usually operated in the swelled (wetted or scoured) condition which implicates the process with aqueous diffusion. As we shall see, the multicomponent diffusion formalism is not affected by these distinctions, but the atomistic interpretation is necessarily different.

Diffusion in zeolite channels has been most successfully dealt with as a surface-moderated diffusion process in which the specie i gas pressure in the channel has the same chemical potential according to the Langmuir isotherm as that of the i specie adsorbed on the channel surface (2,3). Clearly, as long as the channel width is much less than the gaseous mean free path (far below the Knudsen limit) there will be a unique tracer coefficient D_i which is dependent on channel density and cross-section. The question as to whether p_i is actually defined in the channel hinges on the reasonable assumption that the kinetic energy partitions uniformly across the channel surface.

A notable feature of this dry structure is that a binary gas exhibits an independent 2x2 diffusion matrix, the effective solvent being vacant sites on the channel wall, so the Onsager relations

$$J_1 = - L_{11} \nabla \mu_1 - L_{12} \nabla \mu_2 \tag{1}$$

and

$$J_2 = - L_{21} \nabla \mu_1 - L_{22} \nabla \mu_2 \tag{2}$$

with $L_{12} = L_{21}$ are non-trivial. For the usually assumed Henrian dilute solution on the surface

$$\mu_i = \mu_{io} + RT\, \ell n\, p_i(X_i) \tag{3}$$

where the X_i are mole fractions. Thus for a binary gas we obtain with Karger and Bulow (3; KB) the ternary equations

$$\frac{\partial X_1}{\partial t} = \nabla (D_{11} \nabla X_1) + \nabla (D_{12} \nabla X_2) \tag{4}$$

and

$$\frac{\partial X_2}{\partial t} = \nabla (D_{21} \nabla X_1) + \nabla (D_{22} \nabla X_2) \tag{5}$$

absorbing the usual complications associated with the varying molar volumes into the D-matrix. The latter is subject to

$$D_{ij} = RT \left(L_{ii} \frac{\partial \ell n\, p_i}{\partial X_i} + L_{ij} \frac{\partial \ell n\, p_j}{\partial X_i} \right) ; \tag{6}$$

Now by analogy with the interstitial case, or simply according to diluteness, we can proceed with the Darken-LeClaire assumption (4,5)

$$L_{12} = L_{21} = 0 \quad ; \quad D_i = RT\, L_{ii}/X_i \tag{7}$$

which can be rigorously justified for interstitial diffusion and as usual Eq. 6 becomes

$$D_{ij} = D_i \frac{X_i}{X_j} \frac{\partial(\ell n\, p_i)}{\partial(\ell n\, X_i)} \tag{8}$$

Proceeding with KB (3) we explicitly introduce the Langmuir adsorption isotherm so as to eliminate the unmeasurable pressure from the equations. For the binary system this is approximated by the ideal proportionation

$$X_i = \frac{X_m b_i p_i}{1 + b_1 p_1 + b_2 p_2} \tag{9}$$

where the b's are relative sticking probabilities and X_m is the fractional saturable surface sites, assumed to be the same for both species.

It is mathematically convenient to examine the case

$$D_1 >> D_2 \tag{10}$$

which further associates the development with the ternary interstitial volume diffusion case. This will be applied to the solution of the diffusion equations wherein the matrix from Eqs. (8) and (9) is

$$D_{11} = D_1 \frac{X_m - X_2}{X_m - X_1 - X_2} \quad ; \quad D_{12} = D_1 \frac{X_1}{X_m - X_1 - X_2} \tag{11}$$

$$D_{21} = D_2 \frac{X_2}{X_m - X_1 - X_2} \quad ; \quad D_{22} = D_2 \frac{X_m - X_1}{X_m - X_1 - X_2} \tag{12}$$

These define the determinant

$$D = D_1 D_2 \frac{X_m}{X_m - X_2 - X_1} = D_1 D_2 (1 + b_1 p_1 + b_2 p_2) > 0 \tag{13}$$

and the trace

$$D_{11} + D_{22} > 0 \tag{14}$$

so the diffusion solutions are always stable (6). The ratio

$$\frac{D_{12}}{D_{11}} = \frac{X_1}{X_m - X_2} \tag{15}$$

which is the all-important quantity for the diffusion interactions in case (10), can become much greater than unity if p_2 is sufficiently increased. From Eq. (15) it can be seen that this trend will be abetted by a low saturation concentration of adsorbed sites X_m.

With Karger et al. (2,3) we will identify n-heptane with component 1 and benzine with component 2 and note that empirically relation (10) obtains. Assuming with KB that the microporous crystals of NaX approximate to spheres the differential equations become

$$\frac{\partial X_1}{\partial t} = \frac{1}{r^2} \frac{\partial}{\partial r} \left[r^2 \left(D_{11} \frac{\partial X_1}{\partial r} + D_{12} \frac{\partial X_2}{\partial r} \right) \right] \tag{16}$$

$$\frac{\partial X_2}{\partial t} = \frac{1}{r^2} \frac{\partial}{\partial r} \left[r^2 \left(D_{21} \frac{\partial X_1}{\partial r} + D_{22} \frac{\partial X_2}{\partial r} \right) \right] \tag{17}$$

KB proceed by introducing the quasi-steady approximation ($\partial X_1/\partial t = 0$), but this destroys the initial transient which Fig. 2 shows is all-important. The appropriate initial approximation arises from the fact that the kinetics of the slow diffuser should be only marginally affected by the fast diffuser (6) so Eq. (11) can be approximated by

$$\frac{\partial X_2}{\partial t} \simeq \frac{D_{22}}{r^2} \frac{\partial}{\partial r} r^2 \frac{\partial X_2}{\partial r} \tag{18}$$

During the initial transient component 2 sees an approximate one-dimensional problem with the solution

$$X_2 = X_2^0 (1 - \operatorname{erf} r/2\sqrt{D_{22} t}) \tag{19}$$

Further, to keep the problem analytic, we make the same assumption for component 1 which ultimately leads to error in quantity but not in principle. Setting D_{12} equal to its average value (D_{11} and D_{22} should be approximately constant) we obtain the well-known solution (6)

$$X_1 = X_1^0 + \frac{\overline{D}_{12}}{D_{11}} X_2^0 \operatorname{erf}(r/2\sqrt{D_{22} t}) - \left[\frac{\overline{D}_{12}}{D_{11}} X_2^0 + X_1^0\right] \operatorname{erf}(r/2\sqrt{D_{11} t}) \tag{20}$$

To proceed further we require representative diffusion parameters which can only be extracted from integral data like Fig. 2 (2). If in respect to this data we estimate that the beginning of reversion of component 1 at ~900 s occurs with deep impingement at the particle center and saturation at ~18,000 s occurs with deep impingement of component 2 then we can write

$$\sqrt{D_{11}(900)} \simeq \sqrt{D_{22}(18{,}000)} \tag{21}$$

or

$$D_{11}/D_{22} \simeq 20 \tag{22}$$

Arguing along the same lines from the relative magnitudes of integral X_1 and X_2 at the point of reversion in a sphere (~10/1) and integrating the quasi-steady approximation from Eq. 16 (valid for later times)

$$\frac{\partial X_1}{\partial r} \simeq -\frac{\overline{D}_{12}}{D_{11}} \frac{\partial X_2}{\partial r} \tag{23}$$

we can estimate (16)

$$D_{12}/D_{11} \sim 3 - 10 \tag{24}$$

which is very large but potentially consistent with Eq. (15). Taking the boundary condition as saturation values from Fig. 2 we can generate Fig. 3 from estimates 20 and 24 and Eq. (20) for the early part of the transient. Projecting this solution qualitatively to longer times and spherical geometry with impingement as in Fig. 4 one can see that the reconstruction is consistent with the integral data of Fig. 2.

We have made several references to the interstitial volume diffusion case as possibly analogous. Here, where the solutions are always dilute we have (6)

$$\frac{D_{12}}{D_{11}} \simeq \varepsilon_{12} X_1 \;;\; \frac{D_{21}}{D_{22}} = \varepsilon_{21} X_2 \;;\; \varepsilon_{12} = \varepsilon_{21} \tag{25}$$

and these are to be compared in the dilute limit with

$$\frac{D_{12}}{D_{11}} = \left(\frac{1}{X_m}\right) X_1 \;;\; \frac{D_{21}}{D_{11}} = \left(\frac{1}{X_m}\right) X_2 \;;\; \frac{1}{X_m} = \frac{1}{X_m} \tag{26}$$

Furthermore, the thermodynamic interaction parameters have the form e/kT where e is a pair interaction energy. We can speculate that X_m can be expressed in the form kT/e' expressing the proposition that the number of dangling bonds or sites for adsorption in the channels increases with temperature.

The reversion phenomenon, identified here has also been observed on a much larger scale in metal-slag reactions where the latter sometimes polymeric phase provides channels which accommodate fast-moving ions injected from the metal phase. In a later section we will explore the analogy more fully, but first we will examine another gel penetration reaction which may bear on the synthesis.

Mixture Dyeing of Cellophane

Sekido and Morita have recorded a data set for the adsorption and cocurrent diffusion of the dyes Chrysophenine G (dye 1) and Serius Red 4-B* (dye 2) into swelled cellophane. Their matrix of average diffusion coefficients as a function of dye concentration (C_1, C_2) is given in Table 1.

TABLE 1
Mixture Dyeing at 90°C

C_1 (10^4mol/ℓ)	C_2 (10^4mol/ℓ)	D_{11} (10^9mm^2/s)	D_{12} (10^9mm^2/s)	D_{21} (10^9mm^2/s)	D_{22} (10^9mm^2/s)
0.94	0.63	6.5	3.0	−0.012	5.2
1.02	0.82	6.5	2.8	−0.049	5.7
1.08	1.04	6.5	2.5	−0.007	6.1
1.14	1.27	6.5	2.3	−0.002	6.2
1.20	1.52	6.5	2.1	−0.001	6.3
1.17	0.51	7.2	3.5	−0.44	4.7
1.45	0.56	7.8	3.9	−0.61	4.7
1.59	0.57	8.4	4.5	−0.75	4.7
1.72	0.60	8.9	5.0	−0.88	4.7
1.97	0.62	9.6	5.6	−1.07	4.7
0.88	0.46	6.5	3.2	−0.27	4.6
1.26	0.72	7.2	3.4	−0.23	5.4
1.53	0.88	7.6	3.5	−0.27	5.6
2.31	1.39	8.9	4.2	−0.45	6.1
2.82	1.75	9.8	5.1	−0.55	6.4

The data set is reasonably consistent in that the on-diagonal D's are relatively independent of concentration, magnitudes of the off-diagonal D's tend to increase linearly with concentration and are quite large relative to those obtained in dilute liquid diffusion. The last element suggests a strong internal adsorption component as in the zeolite case. However, the change in the sign from D_{12} to D_{21} seems anomalous. Almost any thermodynamics based theory generates a Maxwell relation requiring the same sign of interaction (cf. the examples of the interstitials and zeolites discussed above). However, if there is a Kirkendall type convective backflow of water from the channels by displacement which necessarily correlates the motion of components 1 and 2 then there exist ideal relations of the form (6)

$$D^k_{ii} = (X_j + X_j)\, D_i + X_i\, D_k \tag{27}$$

and

$$D^k_{ij} = -\, X_i(D_j - D_k) \tag{28}$$

where the D's are tracer coefficients and k is the solvent water. Thus D_{12} and D_{21} in a simple aqueous environment can have different signs depending on the magnitudes of D_1 and D_2 relative to D_{H_2O}. If this kinetic effect is combined with the symmetric adsorption effect which necessarily affects the tracer coefficients then one of the off-diagonal coefficients could be augmented and the other reduced to zero, or undergo a sign change. Both eventualities are observed in Table 1. One may expect this unasymmetric behaviour to be common to wet zeolites and swelled polymers.

Reversion in Slag-Metal Systems

A liquid metal-slag reaction has much in common with the gas-zeolite reaction. Noteworthy is their common use as a phase purifier. A typical metallurgical slag consists of a molten mixture of oxides such as C_aO-Al_2O_3-SiO_2 which in the high SiO_2 (acid) range is partly polymerized due to the ordering properties of the silicate SiO_4 tetrahedra and therefore sustains a rather open structure like a zeolite. They are high temperature (~3,000°C; kT ~ 0.5 e.v.) ionic structures exhibiting ionic conductivity favouring cation diffusion. In contact with a molten alloy (e.g. Mn, S, P, and Si in Fe or blast furnace iron) which may be thought of as highly ionized, the slag will be seen by most of the alloying elements as a low chemical potential sink and will diffuse as highly mobile cations into the slag, carrying sufficient electrons to maintain neutrality. The adsorption theory presented above for zeolites would appear to be highly relevant. Since the mobility of alloying element cations can be widely different, diffusion interactions leading to reversion are a distinct possibility.

While practical slag-metal systems are finite they are large enough to appear infinite under quiescent conditions. However, there is always natural or imposed convection so the diffusion controlled region exists only near the surface-tension stabilized interface. One can therefore model an active n-component system with quasi-steady diffusion across a boundary layer in the slag whose width is determined by the convection rate. This arrangement sustains the propensity for the reversion phenomenon for the fast diffuser but somewhat weakened by the necessary convective mixing of all species. One of us proposed a similar model (8) to explain the experimental reversion phenomenon observed by Ramachandran et al. in 1956 (9). However, since the interactions were then conceived to be of the solution type in the metal the magnitudes were not tenable. In the polymeric model where the concentration of adsorption sites X_m may be quite small, the dissolved cation concentration can approach X_m and the cross-interaction ratio D_{ik}/D_{ii} become very large (Eq. (15)) by analogy with the NaX zeolite examined above. We suggest that future investigations of slag-metal reaction rate chemistry should be broadened from the conventional interface rate control models to include diffusion interactions as an explicit and significant component. Further insights are to be gained from theories of membrane diffusion which are also open systems processing very large diffusion interactions (6).

Acknowledgement

The authors are grateful to Professor W-K. Lu for some comments on slag structure.

References

1. R.M. Barrer, Ber. Bunsenges 69, 286 (1965).
2. J. Karger, M. Bulow and W. Schirmer, Z. Phys. Chemie, Leipzig 256, 144 (1975).
3. J. Karger and M. Burlow, Chem. Eng. Sci. 30, 893 (1975).
4. A.D. LeClaire, Prog. Met. Phys. 4, 265 (1953).
5. L.S. Darken, in Atom Movements, ed. J.H. Holloman, ASM Cleveland (1951).
6. J.S. Kirkaldy and D.J. Young, Diffusion in the Condensed State, pp. 96, 106, 156, 163, 334, 336, 338, Institute of Metals, London (1987).
7. M. Sekido and Z. Morita, Bull. Chem. Soc. Jap. 36, 1602 (1963).
8. J.S Kirkaldy, Steelmaking: The Chipman Conference, p. 124, The MIT press, 1965.
9. S. Ramachandran, T.B. King and N.J. Grant, Trans. Met. Soc. 206, 1549 (1956).

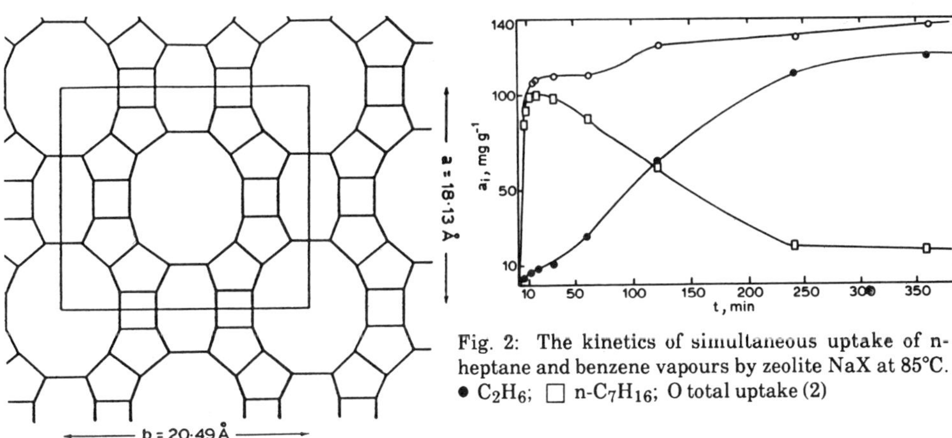

Fig. 1: The zeolite mordenite in cross-section normal to wide channels (1).

Fig. 2: The kinetics of simultaneous uptake of n-heptane and benzene vapours by zeolite NaX at 85°C. ● C_2H_6; □ n-C_7H_{16}; O total uptake (2)

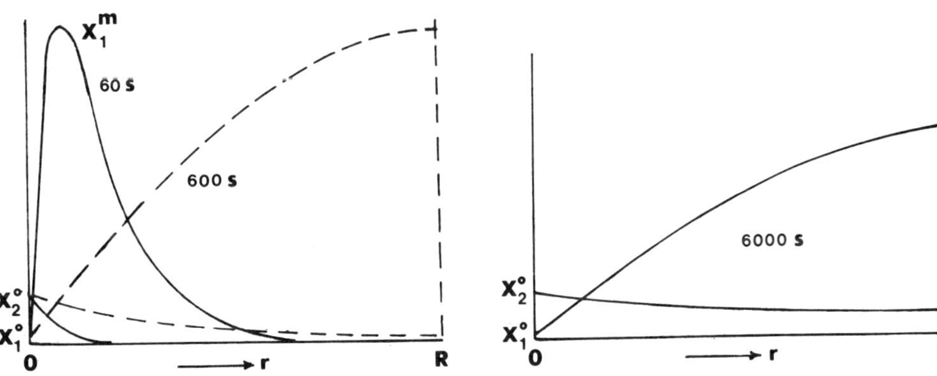

Fig. 3: Semi-quantitative diffusion profiles in zeolite Na-X corresponding to early transient. Fast diffuser n-$C_7H_{16} = 1$.

Fig. 4: Semi-quantitative diffusion profiles in zeolite Na-X corresponding to approach to equilibrium.

Utilization of computer modelling in improving the endothermic gas case carburization of gear steels

R.Wm. Larson
Eaton Corporation, Kalamazoo, Michigan 49003, U.S.A.

Abstract

Case carburization has been effectively simulated by a simple boost-diffuse computer model. this model is coupled with a Lamont Transform based on the Jominy equivalent hardness prediction technique to calculate effective case depth. J. Kirkaldy originally developed this program and it has since been modified by Eaton Corporation. Both the carburization and hardenability models are special topic problems in ternary diffusion. Additional consideration has been given to two additional ternary problems of retained austenite and selective alloy oxidation (grain boundary oxidation).

This program has been used as an aid to improve the endothermic gas carburization of gear steels. This assistance has largely been based on simulated experiments and improved understanding of the process. These simulated experiments include:

- Key variable identification and parametric sensitivity analysis which for all gear steels studied indicated that the most significant variables were hardenability, boost temperature, and boost carbon potential.
- Material specification and process capability.
- Multiple material compatibility under similar operating conditions.
- Hybridization usage of output of program in other programs involved with residual stress development and gear performance.
- designed experiments to explore process optimization.

Computer simulation of multicomponent diffusional transformations in steel

J.-O. Andersson*, L. Höglund, B. Jönsson and J. Ågren
Division of Physical Metallurgy, Royal Institute of Technology, S-100 44 Stockholm, Sweden

ABSTRACT
The thermodynamic and kinetic basis for simulation of multicomponent diffusional transformations is reviewed. The concepts underlying a new program package called *DICTRA* are described. The new software utilizes a new numerical procedure for solving a system of coupled diffusion equations and is interfaced with the THERMO-CALC system for calculation of local equilibrium at a moving phase interface.

The DICTRA program is now applied to diffusional transformations in steel and some recent results will be presented.

1. INTRODUCTION

Calculations of diffusional transformations in multicomponent systems usually involve tedious numerical computations. A realistic treatment of the diffusion requires the solution of a system of partial differential equations and the composition at the moving phase interface needed as boundary condition must be fixed by some method. It is common to assume that equilibrium prevails locally at the phase interface, see ref. 1, and in a binary alloy the composition can be read directly from the phase diagram for the temperature under consideration. In a multicomponent system the situation is more complex because the phase diagram is multidimensional and for a given temperature there are an infinite number of possible equilibrium compositions at the phase interface.

A decade ago one of the present authors (JÅ) presented a numerical method for simulating diffusional transformations under local equilibrium. The method was based on a numerical solution of the multicomponent diffusion equations and calculation of the local equilibrium at a moving phase interface (2). Over the years the program was successfully applied to a number of practical problems, see for example refs. 3 and 4. However, in some cases excessive computer time was needed and numerical performance was less satisfactory.

More recently a new method for solving a system of coupled diffusion equations was suggested by Ågren (5) and a Fortran subroutine package was implemented on the VAX computer. Moreover, the software for calculating equilibrium in multicomponent systems, which was given the name THERMO-CALC (6), improved considerably during the last decade. Thus, it was tempting to combine these two program packages in order to obtain a new improved package for simulation of diffusional reactions in multicomponent systems.

*Present address: Avesta AB, S-774 01 Avesta, Sweden

In the present report we will first describe the basic mathematical models for diffusional transformations in multicomponent alloys and the construction of a software package which was given the name DICTRA. We will then turn to some examples of practical calculations with the new software.

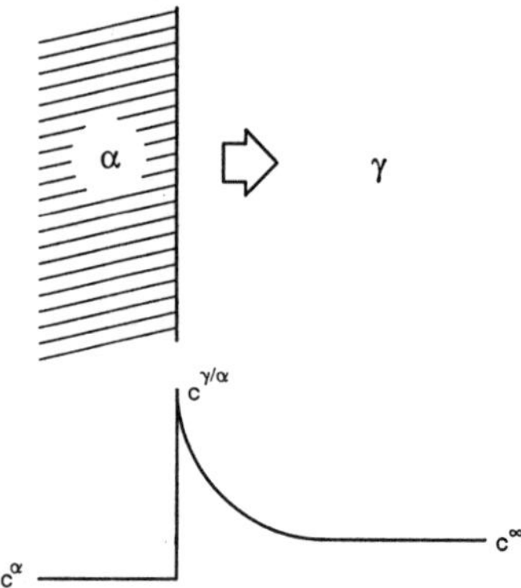

Fig. 1: α phase particle growing into γ. The corresponding concentration profile is shown in lower part of the figure.

2. MATHEMATICAL MODELS

2.1 FLUX BALANCES

Consider a single α particle growing in a supersaturated γ matrix, see Fig. 1. Since the two phases have different compositions the phase transformation must be accompanied by some diffusion in either one or both of the two phases. In fact, in a system with n components the following n balance equations must be obeyed

$$\upsilon^{\alpha} c_k^{\alpha} - \upsilon^{\gamma} c_k^{\gamma} = J_k^{\alpha} - J_k^{\gamma} \qquad k = 1, 2 \ldots n \qquad (1)$$

where υ^{α} and υ^{γ} are the interface migration rate as measured in the local frame of reference in the α and γ phase respectively. c_k^{α} and c_k^{γ} are the concentrations (mole per unit volume) of component k in both phases close to the phase inteface and J_k^{α} and J_k^{γ} the diffusive fluxes on each side of the interface. As a local frame of reference it is convenient to apply the volume fixed frame of reference which is defined from the condition

$$\sum_{k=1}^{n} J_k^{\alpha} V_k^{\alpha} = 0 \qquad (2)$$

and a similar expression in the γ phase. V_k^{α} is the partial molar volume of component k in the α phase. It is always possible to derive a relation between υ^{α} and υ^{γ} and eliminate either one of them and one of the equations. For example, consider as an approximation the case where all the substitutional elements of a phase have the same partial molar volume, say V_S^{α} in α and V_S^{γ} in γ and further the two contain the same substitutional elements. Then we have

$$\upsilon^{\alpha} = \upsilon^{\gamma} V_s^{\alpha} / V_s^{\gamma} \qquad (3)$$

It should be emphasized that the two phases may have different molar volumes and the transformation may thus result in a volume change even though the diffusion causes no change of volume. On the other hand, if the γ phase consists of only components which are interstitial in α, e.g. carbon in graphite, then we have instead

$$\upsilon^{\alpha} = 0 \tag{4}$$

All the balance equations for the substitutional components in α are then eliminated and only the ones for the interstitial components remain. The volume change caused by the reaction is

$$\frac{1}{A}\frac{dV}{dt} = \upsilon^{\alpha} - \upsilon^{\alpha} \tag{5}$$

where A is the area of the moving phase interface. It is convenient to introduce a new concentration variable u_k related to c_k by means of

$$c_k^{\alpha} = \frac{u_k^{\alpha}}{V_s^{\alpha}} \quad \text{i.e.} \quad u_k^{\alpha} = \frac{x_k^{\alpha}}{\sum x_j^{\alpha}} \tag{6,7}$$

where x_k^{α} is the mole fraction of component k. The summation in the denominator is only performed over the substitutional components.

2.2 LOCAL EQUILIBRIUM

In the preceeding section we found that we have n-1 flux balance equations in an n-component system. These equations have to be solved simultaneously. The rate υ^{α} or υ^{γ} is then

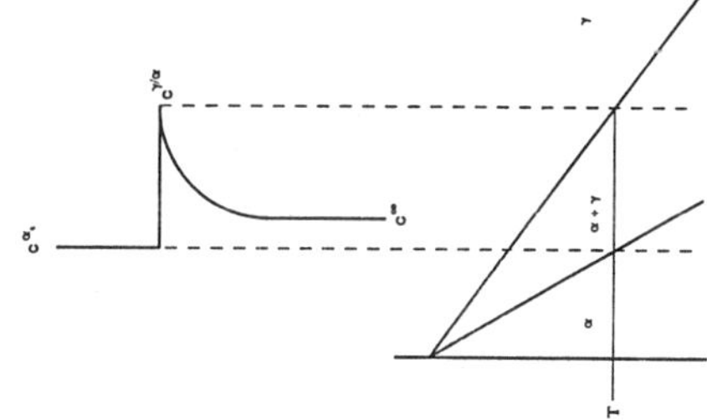

Fig. 2: Graphic construction to get the concentrations close to phase interface from binary phase diagram when local equilibrium is established at the interface. The temperature under consideration is T'.

determined and the rate which was eliminated is subsequently obtained from the relation between them. If there are n-1 equations there must also be n-1 unknowns in order to pose a well-defined problem with a unique solution. In a binary system n=2 and there is only one equation. Assuming that equilibrium prevails locally at the phase interface the compositions c_k^{α} and c_k^{γ} can be read directly from the phase diagram for the given temperature and pressure. The single equation can then be used to calculate υ^{α} or υ^{γ}, see Fig. 2. In a system with $n > 2$ there are an infinite number of possible concentrations. However, each combination of possible concentrations is determined by n-2 activities when pressure and temperature are fixed. These activities are unknown and the total number of unknowns, i.e. the rate and the activities, then is

$$1 + n - 2 = n - 1$$

We have thus the same number of unknowns as balance equations and there may be a unique solution to the system of equations represented by eq. 1. In principle one can now guess the unknown activities, calculate the compositions c_k^α and c_k^γ from the guessed activity values and the given pressure and temperature. One can subsequently calculate the diffusive fluxes and then check if all n-1 balance equations are obeyed. If not, one has to guess new activity values and the rate υ by some iteration technique until the equations are obeyed.

In the Thermo-Calc program it is possible to calculate an equilibrium under fixed pressure, temperature and component activities from the thermodynamic functions of the two phases. Thermo-Calc will now be combined with a procedure to solve the diffusion problem and a Newton-Raphson iteration technique to determine the rate and the unknown activities. These matters will be further discussed in section 3.

2.3 DIFFUSION IN THE ONE-PHASE REGIONS

The diffusion inside each phase is determined by the mass-conservation law

$$\frac{\partial c_k}{\partial t} = - \operatorname{div}(J_k) \tag{8}$$

where the fluxes J_k are given by the multicomponent version of Fick's law, i.e.

$$J_k = - \sum_{j=1}^{n-1} D_{kj}^n \nabla c_j \tag{9}$$

The calculation of the diffusion coefficient matrix D_{kj}^n is described in detail in the report by Andersson and Ågren (7) and will not be discussed further here. We will only emphasize that this calculation involves both thermodynamic and kinetic information and in general it leads to concentration dependent diffusivities. For each time step Δt Eq, 8 is solved by means of the numerical method developed by Ågren (5). The magnitude of the time step will be controlled by an automatic procedure described in ref. 8.

3. CONSTRUCTION OF THE COMPUTER PROGRAM

The computer program will consist of essentially 4 modules, namely

. Displacement of the phase interface
. Solution of the flux-balance equations, eq. 1
. Calculation of two-phase equilibrium composition
. Solution of diffusion equations, eqs. 8 and 9

The program was given the name *DICTRA* (Diffusion Controlled Transformations). The routine for solving the diffusion equations was taken directly from Ågren (5) and a routine from the Harwell Subroutine library (9), NS01A was applied for solving the system of flux balances. The Thermo-Calc program package (6) was applied for the equilibrium calculations. A separate procedure was developed in order to displace the phase interface and perform various actions connected with this displacement. The procedure is based upon the method suggested by Murray and Landis (10) although important improvements are introduced. See the report by Crusius et al. (11).

In addition a user interface was constructed. The user interface is the interactive part of the program which allows the user to define what actions the program should take. It consists of a set of commands and a procedure that translates the user input given into suitable actions. All the data used in the program and generated by the program must be stored in some suitable

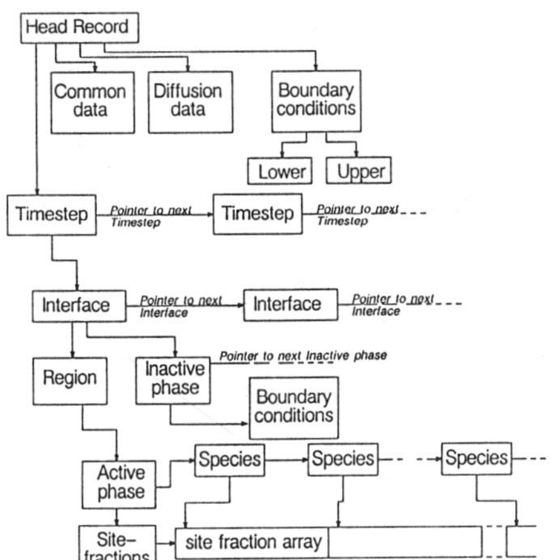

Fig. 3: Overview over data structure.

way. In the present case the method of linked lists was applied. A schematic representation of the data structure used is shown in Fig. 3.

It should be emphasized that the problem of constructing a user interface and a convenient management of the data structure is quite tedious from a programming point of view. In the present project it was possible within a rather short time to solve this problem only because a subroutine package developed by Sundman (12) was used.

A special module was constructed in order to handle a databank with kinetic information. The concentration, temperature and pressure dependence of these quantities were represented with the same type of series expansion that are used when storing thermodynamic data. It was thus possible to apply rather directly programs that were already available for handling thermodynamic databanks.

4. REVIEW OF PREVIOUS CALCULATIONS

The diffusional growth or shrinkage of particles has been treated numerically by several authors. Especially in solidification, where the mathematics is analogous although one considers heat flow rather than diffusion, there is a rich litterature which will not be considered here. Among the early works on diffusion-controlled solid state transformations the work by Goldstein and Ogilvie (13) and by Tanzilli and Heckel (14,15) deserve mentioning. Although they did not consider multicomponent alloys they developed numerical procedures similar to the one of the present report and applied the equilibrium phase diagram to evaluate the composition at the phase interface. The method by Tanzilli and Heckel was later applied to various practical situations, see for example ref. 16 where heat treatment of α and β brass diffusion couples were simulated. Randich and Goldstein (17) went one step further and developed a method applicable for ternary systems. Their method was applied to the growth of phosphide in Fe-Ni-P alloys and to the α - γ transformation in Fe-Ni-Co alloys (18). The local equilibrium at the phase interface was obtained from a simple mathematical representation of the ternary phase diagrams rather than a thermodynamic calculation. This approach is more difficult to extend to higher order systems than the thermodynamic one described in the present report. However, they were able to predict many aspects of the complex multicomponent behaviour, e.g. the shift in interface composition with time caused by different magnitudes of the diffusion coefficients in combination with a finite size of the system.

To the knowledge of the present authors Ågren (2) was the first one who developed a method that combined a numerical solution of the diffusion equations with a thermodynamic calculation of the driving force for diffusion as well as the local equilibrium at the phase interface. The method was first applied to the α - γ transformation during intercritical annealing of Fe-Mn-C dual phase steels as well as the subsequent cooling (3) and later to a number of practical problems. In Fig. 4 from ref. 3 the calculated Mn and C content in austenite close to the austenite/ferrite phase interface are plotted as functions of time during isothermal annealing at 740 °C. As can be seen, there is an enrichement of Mn in the austenite.

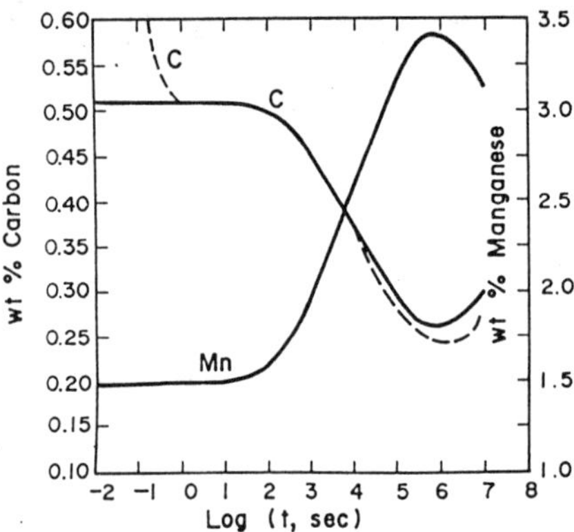

Fig. 4: The calculated variation of manganese and carbon content in the austenite close to the phase interface as a function of time at 740 °C. The dashed line shows the variation of the carbon content at the center of the austenite region. From ref. 3.

It is interesting to notice the predicted difference in carbon content between center and phase interface caused by the thermodynamic coupling between the C and Mn diffusion. .

5. SOME APPLICATIONS OF DICTRA

5.1 DISSOLUTION OF CEMENTITE IN Fe-Cr-C

The DICTRA program has been applied to the dissolution of cementite in a ternary Fe-2.06 at-%Cr-3.91 at-%C alloy at 910 °C. The calculations will be compared with recent experimental information by Liu and Ågren (20). The calculations are described in more detail in ref. 21.

As initial compositions of the cementite and the austenite those inherited from the cementite and ferrite, respectively, after the soft annealing at 735 °C, were taken. It was thus assumed that the ferrite matrix transforms into austenite instantly before there is any cementite dissolution. This also fixes the initial mole fraction of cementite. It was assumed that the spheroidal particles maintain their shape throughout their dissolution and just undergo a gradual decrease in size with time. For the setup of the simulation, the thickness of the austenite shell surrounding the cementite was choosen to make the average composition of the simulated system equal to the alloy composition.

The thermodynamic data from Hillert and Qiu (22) was applied in the simulation and the kinetic parameters were taken from the assessment by Fridberg et al. (23) and corrected with the thermodynamic factor. The diffusivity of Cr in cementite is not known and different values were tried until a reasonable agreement between experiments and calculations was obtained. A value of about 1/40 :th of the value reported for the diffusivity of Cr in fcc Fe (23) was finally chosen.

Figs. 5a-c show the calculated Cr concentration profiles in comparison with the experimental ones determined by Liu and Ågren for 10, 100 and 10000 s austenitizing time. As can be seen, the agreement is quite good. In Fig. 5d we compare our calculated volume fraction of cementite with the experimental one. Also here the agreement is good. The dashed line is the measured volume fraction of cementite in the pre-annealed material.

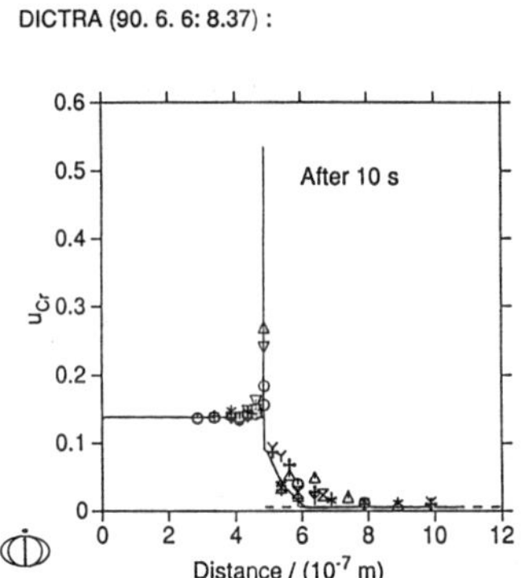

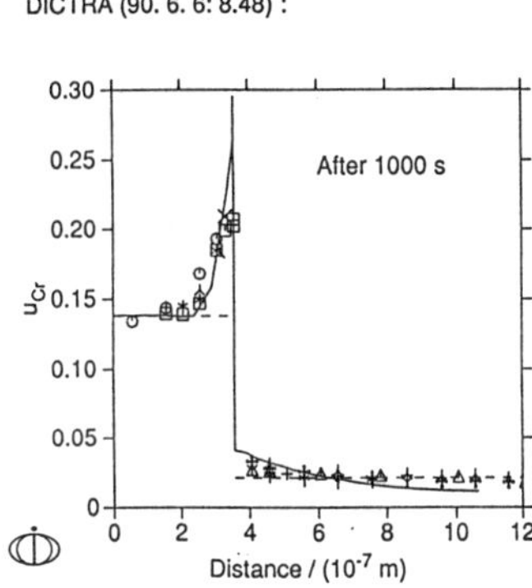

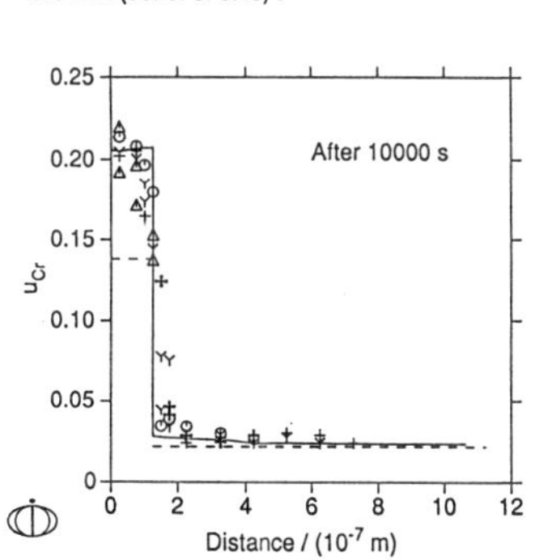

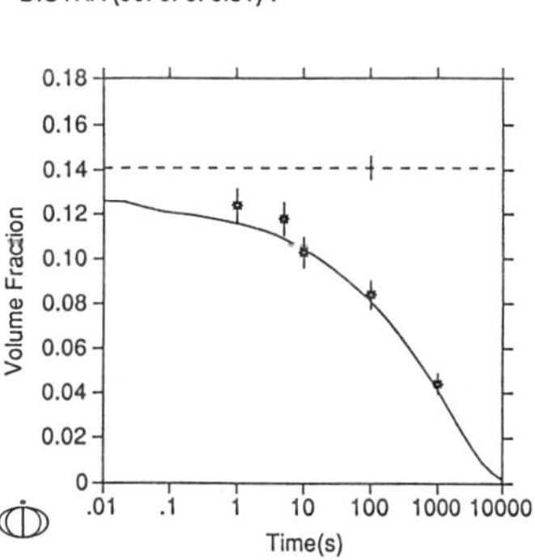

Fig. 5: Experimental and calculated Cr-concentration profile in cementite and γ after a) 10 s b) 1000 s and c) 10000 s at 910 °C. d) Experimental and calculated volume fraction of cementite as a function of time at 910 °C. From ref. 21.

5.2 THE AUSTENITE - FERRITE TRANSFORMATION IN Cr-Fe-Ni ALLOYS.

In this section we will present simulations of the α/γ transformation in ternary Cr-Fe-Ni α/γ diffusion couples. The simulations will be compared with the experimental work by Im (24) who considered four different diffusion couples A, B, C and D, respectively. Their composition is given in Table 1. All four diffusion couples were annealed at 1100 °C, for various times. The details of the simulations are given in ref. 25.

The thermodynamic description by Hillert and Qiu (26) was applied. For the austenite and ferrite the kinetic parameters were derived from diffusion coefficients reported by Fridberg et al. (23).

Experimental information on ternary diffusion is often represented by means of the so called diffusion-path. Such diagrams are readily plotted using the DICTRA program. The experimental and calculated diffusion paths at various times for couple B are shown in Fig. 6a. The agreement between calculations and experiments may be regarded as good. In particular, one may notice that the calculated tie-lines move in the right direction with increasing annealing time. However, we notice a discrepancy between the calculated and measured composition of the α/α+γ equilibrium for diffusion couple A. This is due to the thermodynamic description used. The thickness of the α layers as function of annealing time is shown in Fig. 6b for all the couples. In A, B, and D the α layer is first growing and later it is shrinking. However, in C it shrinks allready from the start. It is encouraging that the

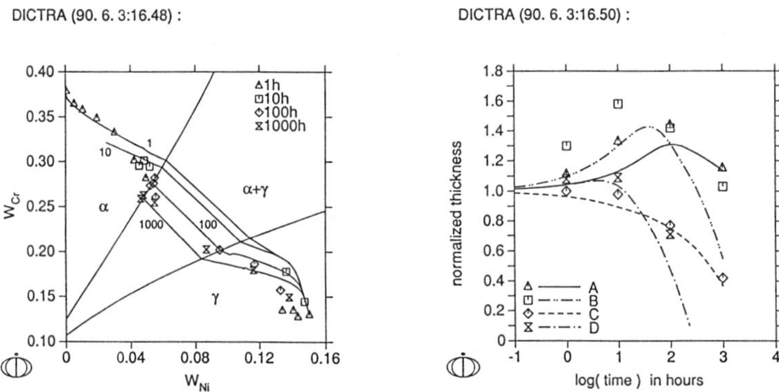

Fig. 6a: Experimental and calculated diffusion paths for diffusion couple B. The different lines are for the various annealing times indicated (1, 10, 100 and 1000h). b) Experimental and calculated thickness of α layer as a function of time. From ref. 25.

simulations correctly describe the general trends even if the absolute values are somewhat off. For all the simulations we notice that for growing α, the calculated growth rate is too low and for shrinking α it is too high. Some examples of experimental and calculated concentration profiles are given in Figs. 7a-d for diffusion couple B after 1, 10, 100 and 1000 h. The α phase is on the left side. It may be noticed that α is growing as long as it has not achieved a homogeneous composition.

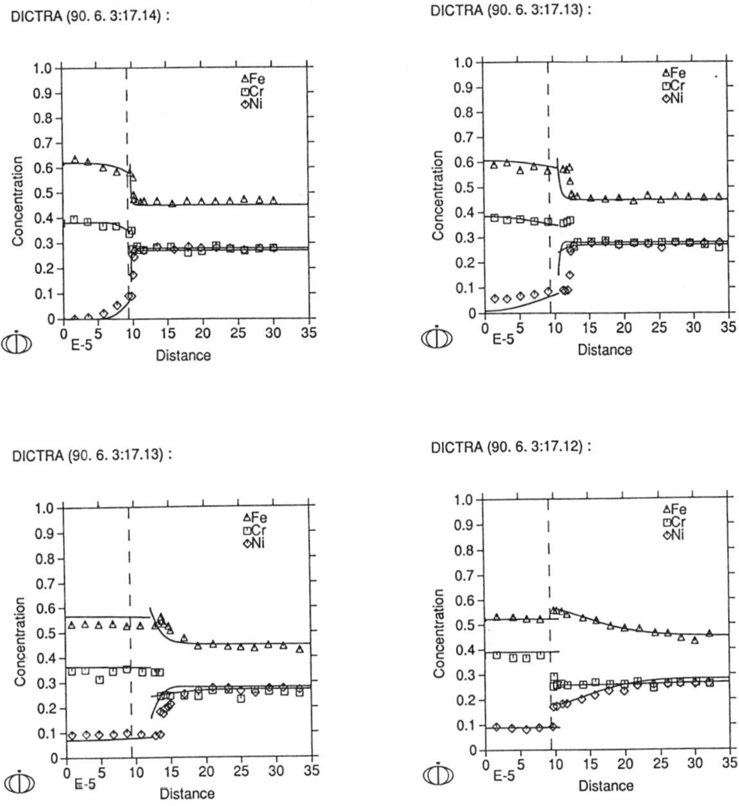

Fig. 7: Experimental and calculated concentration profiles for diffusion couple B after a) 1 h, b) 10 h c) 100 h and d) 1000 h. From ref. 25.

TABLE 1: Composition of duffusion couples

Name	Composition of α phase	Composition of γ phase	Thickness of α phase layer
A:	38 w/o Cr	27 w/o Cr, 28 w/o Ni	186.9 μm
B:	38 w/o Cr	13 w/o Cr, 15 w/o Ni	124.1 μm
C:	37 w/o Cr, 9 w/o Ni	24 w/o Cr, 32 w/o Ni	130.6 μm
D:	37 w/o Cr, 1 w/o Ni	13 w/o Cr, 15 w/o Ni	71.6 μm

We can conclude that the diffusion rates in our simulations have been too low in α and too high in γ. It should be emphasized though, that the kinetic data used is not valid for concentrated alloys. In order to improve the simulations one must compile better kinetic data. Such work is under progress at the Division of Physical Metallurgy.

6. CONCLUSIONS

Indeed, it has been possible to combine numerical calculations of diffusion with refined thermodynamic calculations in order to predict phase transformations of multicomponent alloys from fundamental thermodynamic and kinetic data. Such calculations have a strong predictive power and involve no adjustable parameters. The present examples show that in spite of the lack of accurate kinetic data the results obtained are very encouraging. The DICTRA program package works properly and should be quite a valuable tool provided that proper kinetic and thermodymanic data are available.

ACKNOWLEDGEMENTS

Financial support from the Swedish Board for Technical Development and from the Volkswagen Stiftung in West Germany is gratefully acknowledged.

REFERENCES

1. M. Hillert and J. Ågren: in *Advances in Phase Transitions,* eds. J.D. Embury and G.R. Purdy, Pergamon Press 1988, p 1
2. J. Ågren: J. Phys. Chem. Solids, vol 43 (1982) p 385
3. J. Ågren: Acta Metall. vol 30 (1982) p 841
4. J. Ågren and G.P. Vassilev: Materials Science and Eng., vol 64 (1984) p 95
5. J. Ågren: Internal Report, series D 84, 1987, Division of Physical Metallurgy, Royal Inst. of Technology, S-100 44 Stockholm, SWEDEN
6. B. Sundman, B. Jansson and J-O Andersson: CALPHAD 9 (1985) p 153
7. J. Ågren and J.-O. Andersson:
Models for Numerical Treatment of Multicomponent Diffusion in Simple Phases,
TRITA-MAC-0436 May 1990, Internal Report, Div. of Phys. Metall.,
Royal Inst. Techn. S-100 44 Stockholm SWEDEN
8. G. Dahlquist and Å. Björk: *Numerical Methods* Prentice-Hall, Englewood Cliffs, New Jersey (1974)
9. Harwell Subroutine Library Specification, Computer Science and Systems Division, Harwell Laboratory, Oxfordshire OX110RA, U.K.
10. W.P. Murray and F. Landis: Trans. ASME, vol 81 (1959) p 106
11. S. Crusius, G. Inden, U. Knoop, L. Höglund and J. Ågren:
On the Numerical Treatment of Moving Boundary Problems,
TRITA-MAC-0417 Jan. 1990, Internal Report, Div. of Phys. Metall.,
Royal Inst. Techn. S-100 44 Stockholm SWEDEN
12. B. Sundman: Metlib Subroutine Manual, Division of Physical Metallurgy, Royal Inst. of Technology, S-100 44 Stockholm, SWEDEN
13. J. I. Goldstein and R. E. Ogilvie: Geochimica et Cosmochimica Acta, Vol. 29 (1965) p 893
14. R. A. Tanzilli and R. W. Heckel: Trans. TMS-AIME, Vol. 242 (1968) p 2313
15. R. A. Tanzilli and R. W. Heckel: Metall. Trans. Vol. 1 (1970) p 1863
16. R. W. Heckel, A. J. Hickl, R. J. Zaehring and R. A. Tanzilli: Metall. Trans. Vol. 3 (1972) p 2565.
17. E. Randich and J. I. Goldstein: Metall. Trans. A, Vol. 6A (1975) p 1553
18. A. Moren, E. Randich and J. I. Goldstein: in Proceedings of the 1976
International Conference on Computer Simulation for Materials Applications

Nuclear Metallurgy Vol. 20 part 1 p 221
19. J. Ågren, H. Abe, T. Suzuki and Y. Sakuma: Metall. Trans. A, Vol. 17A (1986) p 617
20. Z.-K. Liu and J. Ågren: Dissolution of Cementite in an Fe-Cr-C Alloy, Part I: Experimental Investigation, TRITA-MAC 0418, Jan. 1990, Internal Report, Div. of Phys. Metall., Royal Inst. Techn. S-100 44 Stockholm SWEDEN
21. L. Höglund, B. Jönsson, Z.-K. Liu and J. Ågren: Dissolution of Cementite in an Fe-Cr-C Alloy, Part II: Numerical Simulation, TRITA-MAC, June 1990, Internal Report, Div. of Phys. Metall., Royal Inst. Techn. S-100 44 Stockholm SWEDEN
22. M. Hillert and C. Qiu: A Thermodynamic Assessment of the Fe-Cr-Ni-C system, TRITA-MAC-0434, May 1990, Internal Report, Div. of Phys. Metall., Royal Inst. Techn. S-100 44 Stockholm SWEDEN
23. J. Fridberg, L.-E. Törndahl and M. Hillert: Jernkont. Ann., 153(1969)263-276
24. C.-B. Im: Diffusion phenomena in $\gamma/\alpha/\gamma$ diffusion couples of Fe-Cr-Ni ternary alloys., Bachelor thesis. Tokyo Institute of Technology, Tokyo 152, Japan 1985 (In Japanese)
25. L. Höglund, B. Jönsson and J. Ågren: Simulation of Diffusional Reactions in Cr-Fe-Ni alloys, TRITA-MAC, June 1990, Internal Report, Div. of Phys. Metall., Royal Inst. Techn. S-100 44 Stockholm SWEDEN
26. M. Hillert and C. Qiu: A Reassessment of the Cr-Fe-Ni system, TRITA-MAC-0412, Oct 1989, Internal Report, Div. of Phys. Metall., Royal Inst. Techn. S-100 44 Stockholm SWEDEN

Preferential concentration and depletion of carbides after annealing of mismatched chromium alloy weldments

B. Buchmayr
Institute for Materials Science and Welding Technology, Graz University of Technology, Austria

J.S. Kirkaldy
Institute for Materials Research, McMaster University, Hamilton, Ontario, Canada, L8S 4M1

Abstract

A ternary diffusion model is proposed which allows for simultaneous ternary diffusion and precipitation on one side and carbon migration and carbide dissolution on the other side of the fusion line between 1%CrMoV and 12%CrMoV weldments. The numerical calculations are able to predict the experimentally measured values and lead to an improved understanding of the complex diffusion processes.

Introduction

In steam power generating plants weldments between different kinds of heat resistant steels are commonly used. It is known that due to the chemical composition difference (especially of the Cr-content), microstructural changes occur during post weld heat treatment and in service, viz., a carbon depletion in the low alloy steel heat-affected zone (HAZ) and a carbide seam in the higher alloyed filler metal. These changes, reviewed in (1,2) determine the mechanical properties and the fracture appearance. An extensive study on the dissimilar weldment between a low alloyed 1%CrMoV cast steel and a martenistic 12%CrMoV steel was recently made by Witwer (3). The objective was to quantify the microstructural changes near the fusion line and to study the effects on the toughness and creep rupture behaviour.

In an earlier contribution (4) the authors attributed the preferential precipitation of carbides on the Cr rich side of a post-annealed weldment between a 1%CrMoV and a 12%CrMoV steel to the negative ternary diffusion interaction of C and Cr. In this paper the authors attempt to further quantify the argument using Kirkaldy's model for simultaneous precipitation and diffusion (5,6) for the high Chromium side, where it is assumed that finely-spaced precipitates of low volume fraction are in local equilibrium according to a solubility product with multicomponent diffusion in the matrix according to a Fick-type equation, and for the low Chromium side an Ostwald Ripening approach for simultaneous precipitate dissolution and diffusion.

The Ternary Diffusion Model

On the high Chromium side the amount of precipitates always increases. Here we base the concentration on the schematic Fe-C(1)-Cr(2) phase diagram of Fig. 1 subject to the solubility product relations

$$C_2 = C_2(C_1) \quad ; \quad \partial C_2 = \partial C_1\,(dC_2/dC_1) = m\partial C_1 \quad ; \quad m<0 \qquad (1)$$

and the partition ratio

$$n = (C_{20} - C_2)/(C_{10} - C_1) > 0 \tag{2}$$

where (C_1, C_2) is the matrix composition. The differential equations applied are

$$(C_{10} - C_1)\frac{\partial p}{\partial t} = -\left(\frac{\partial C_1}{\partial t} + \frac{\partial J_1}{\partial x}\right) = -\frac{\partial C_1}{\partial t} + \frac{\partial}{\partial x} D_{11}\frac{\partial C_1}{\partial x} + \frac{\partial}{\partial x} D_{12}\frac{\partial C_2}{\partial x} \tag{3}$$

and

$$(C_{20} - C_2)\frac{\partial p}{\partial t} = -\left(\frac{\partial C_2}{\partial t} + \frac{\partial J_2}{\partial x}\right) = -\frac{\partial C_2}{\partial t} + \frac{\partial}{\partial x} D_{21}\frac{\partial C_1}{\partial x} + \frac{\partial}{\partial x} D_{22}\frac{\partial C_2}{\partial x} \tag{4}$$

where p is the volume fraction of precipitate. Eliminating the left hand sides of Eqs. 3 and 4, and including Eqs. 1 and 2 the combination can be rearranged to give

$$(n - m)\frac{\partial C_1}{\partial t} = \frac{\partial}{\partial x}(D_{11} + D_{12}m - D_{21} - D_{22}m)\frac{\partial C_1}{\partial x} \approx \frac{\partial}{\partial x}(D_{11} + D_{12}m)\frac{\partial C_1}{\partial x} \quad ; \; D_{11} >> D_{22} \tag{5}$$

or

$$\frac{n-m}{m}\frac{\partial C_2}{\partial t} \approx \frac{\partial}{\partial x}\left(\frac{D_{11} + D_{12}m}{m}\right)\frac{\partial C_2}{\partial x} \tag{6}$$

Since m is negative and

$$D_{12}/D_{11} \approx \varepsilon_C^{Cr} X_C \tag{7}$$

where $\varepsilon_C{}^{Cr}$ is the Wagner interaction parameter in ferrite ($= -72$ according to (7)), the solutions for C_1 and C_2 are always stable and both are not too different from the shape of the error function for infinite boundary conditions. We assume a weldment diffusion couple but no unmixing on either side of the weld at $t=0$ (Fig. 2). The profiles, which are to be matched with the symbols in Fig. 1 are the result of integrating Eqs. (5) and (6) for $t>0$. Note that the profile depth of both components in the matrix (Cr only diffuses locally) are determined by the carbon diffusion coefficients discounted approximately by the ratio $1/(n-m)$. Both D_{12} and m as a product increase the effective value of D_{11}. All of this is in accord with what one would intuitively expect.

The relative volume of precipitate p is obtained via Eq. (4) ($D_{11} >> D_{22}$) as

$$p(x) = p_0(x) + \int_0^t (\partial p/\partial t)\,dt \approx p_0(x) - \int_0^t \frac{1}{(C_{20} - C_2)}\frac{\partial C_2}{\partial t}\,dt \tag{8}$$

which gives the right trend as can be seen from the C_2 profile of Fig. 2. Note that the computational problem involves at least the parallel integration of Eqs. (6) and (8) with the possibility of cross-checking through Eqs. (3) and (5).

On the left hand side, where the amount of precipitates decreases we need to include Ostwald Ripening and use an explicit p-dependent formulation. Based on the Gibbs-Thompson effect we can relate the precipitate radius r to the precipitate mole fraction, p, assuming nucleation site saturation on n available sites per unit volume. This yields

$$p = 4/3\,\pi r^3 n = 4/3\,\pi\, n\left[\frac{2\sigma V}{RT}\right]^3 / \left[\ln\frac{X_2^x X_1^y}{K}\right]^3 \tag{9}$$

For the kinetic relations to follow we require

$$p = -P_1 \frac{\partial C_1}{\partial t} - P_2 \frac{\partial C_2}{\partial t} \tag{10}$$

where

$$P_1 = -\frac{\partial p}{\partial X_1} = \frac{4\pi n x}{X_1}\left[\frac{2\sigma V}{RT}\right]^3 / \left[\ln \frac{X_2^x X_1^y}{K}\right]^4 = 3Py/\left(X_1 \ln \frac{X_2^x X_1^y}{K}\right) \tag{11}$$

and

$$P_2 = -\frac{\partial p}{\partial X_2} = \frac{4\pi n y}{X_2}\left[\frac{2\sigma V}{RT}\right]^3 / \left[\ln \frac{X_2^x X_1^y}{K}\right]^4 = 3Py/\left(X_2 \ln \frac{X_2^x X_1^y}{K}\right) \tag{12}$$

Again assuming an infinitesimal volume fraction of precipitates, equal molar volume of precipitate and matrix and ternary diffusion interactions we can formulate the one-dimensional Fick-type equations similar to Eq. 3 and 4 with modified left hand sides and take into account the relevant solubility product relations.

After rearrangement of terms the final equations are similar to those in Eq. 5 and 6. This analogous formulation is quite helpful to fulfill the mass balance condition at the fusion line

$$D_{eff}^{1}\left[\frac{\partial C_1}{\partial x}\right]_{FL}^{1} = D_{eff}^{r}\left[\frac{\partial C_1}{\partial x}\right]_{FL}^{r} \tag{13}$$

where 1 and r mean left and right hand side, respectively.

Numerical Calculation

Based on the aforementioned equations and relations, a numerical procedure was developed to predict the carbon profile and amount of precipitates on both sides of the fusion line. Thereby it was assumed that at the post weld heat treatment (PWHT) temperature (700°C) a carbide equilibrium is reached, i.e. the carbide type on the high Cr (right hand) side is $M_{23}C_6$ and on the low Cr (left hand) side M_3C is formed. To solve the modified Fick-equation the finite difference method was used taking into account the stability criteria for time increment Δt $(2D_{eff}\Delta t/(\Delta x)^2) < 1)$. The solubility product for $M_{23}C_6$ in ferrite was described by

$$\log(Cr^{23}\,C^{6}) = 5.9 - 7375/T \tag{14}$$

according to (8).

The constants C_{20} and C_{10}- in Eq. 2 (Cr and C content in $M_{23}C_6$) were chosen in accordance to (9) to be 70 and 5.7%, respectively. The coefficient m in Eq. 3 and 4 was held constant at the value −5000. For the alloyed cementite (M_3C) the chemical composition of the carbide was assumed to be $(Fe_{5/6}Cr_{1/6})_3C$, i.e., the value x in Eqs. 9-12 is 0.5 and y=1. Regarding the solubility, measured data published in (10) were used for the calculation. The initial and boundary conditions were written as

$$C_i(x_+, 0) = C_{i0} \qquad C_i(x_-, 0) = C_{i1} \tag{15}$$

$$C_i(x_0, t) = C_{i0} \qquad C_i(-x_0, t) = C_{i1} \tag{16}$$

together with

$$\partial C_i/\partial x = 0 \qquad \text{at} \quad x = \pm x_0 \tag{17}$$

For the carbon diffusion coefficient the following values have been used ($D_0 = 2\ mm^2/s$, $Q = 84.1$ kJ/mol) (11). For comparison the values for chromium are $D_0 = 852$ and $Q = 251$ kJ/mol (12), i.e., there is negligible chromium diffusion at the annealing temperature considered.

The calculation procedure comprises the following steps:

- initialization of initial conditions
- determination of Cr and C in solution
- calculation of effective diffusion coefficient and check of stability criteria with respect to each time increment
- solving the modified Fick-type differential equations for $\partial C_1/\partial t$ for both sides
- using the solubility product to determine C_2
- calculation of dissolution rate on the left hand side
- determination of total C-content and total Cr (=const).
- calculation of amount of precipitates
- taking into account the boundary conditions
- setting the input value for the next time step to the previous results.

Results

The solution of the differential equation system with respect to the carbon in solution is shown in Fig. 3a and that for the dissolved chromium is shown in Fig. 3b. The profiles are plotted for an annealing temperature of 700°C after different times (5, 10, 15,. . . .60 min). The profile of the total carbon concentration is shown in Fig. 4. The corresponding amount of precipitates is illustrated in Fig. 5.

At the beginning the carbon migration is driven by the difference in thermodynamic activity caused by the difference in chromium content. On the left hand side carbon depletion takes place combined with the dissolution of M_3C particles. The dissolution rate is given by the difference between the equilibrium concentration at the α/M_3C phase boundary and the level of carbon in solution in the matrix. On the high chromium side, the carbon peak increases up to a saturation level, which is given by the expression

$$\Delta C = \varepsilon_C^{Cr}\, X_C\, \Delta Cr \tag{18}$$

In our case the C peak reaches a level of about 0.75% after one hour of annealing. The time dependence of the development of the carbon peak is shown in Fig. 6, which can be described by a typical $\sqrt{Dt}$ expression.

As soon as the activity difference directly at the fusion line is almost balanced, the carbon enrichment progresses into the high chromium region, whereby the migration speed is reduced by the decreasing driving force caused by the decreasing ternary diffusion effect. Therefore the C-profile becomes more trapezoidal than triangular, so that after long annealing times an almost equally-dense carbide seam is formed. The inability of chromium to diffuse is responsible for this phenomenon occurring, otherwise a common error function profile may be expected. The distribution of chromium in solution is given simply by considering the solubility product.

The distribution of C and Cr in solution determines directly the amount of precipitates, as shown in Fig. 5.

To verify the numerical calculations, a comparison is made with measured C-profiles and investigations of the microstructure with special emphasis on the carbide density (3).

Experimental Verification

Fig. 7 shows an overview of the microstructure at the fusion line after PWHT at 730°C for 2h. As mentioned earlier there is a very dense carbide seam along the fusion line in the higher Cr-alloyed material, whereas in the lower Cr region a carbon depleted zone is visible, which is about twice as broad as the carbide seam. An REM micrograph is shown in Fig. 8 demonstrating the interface region.

Fig. 9 shows carbon scans across the transition zone. The average profile has the same form as predicted for long annealing times. The maximum carbon concentration was found at a level as predicted by the numerical model. In addition, the C-profile in the region of carbide dissolution is close to the predicted one. The absolute values correspond surprisingly well with the numerical results. The thickening of the dense carbide seam as well as of the carbon depleted seam follow the well-known $\sqrt{Dt}$ law.

The consequences of such microstructural changes are obvious and described in detail in (3). Pronounced effects were found with respect to the toughness (the carbide seam determines the crack path) and the creep rupture behaviour (lower creep strength in the C-depleted zone).

Ways of Reducing Carbide Seam Formation

The carbon enrichment or the carbon depletion and the carbide density, respectively, is determined by the difference in the chemical potential of carbon due to differing chromium contents. To characterize the influence of alloying elements on the C-activity the Wagner interaction parameter can be applied to predict the maximum C-peak in the carburized zone. The difference in the thermodynamic potential can be lowered either by alloying carbide forming elements to the lower alloyed steel, by reducing this value in the high alloyed steel or by adding γ-forming elements with positive ε-value to the higher Cr-containing material.

From the practical point of view the following solutions should be applicable for reducing the carbide seam formation

- use of different Cr-alloyed consumables in an appropriate sequence (2 1/4Cr/5Cr/9Cr/12Cr)
- addition of elements of different C-affinity to the lower alloyed steel (i.e. V,Nb,Ti)
- lowering the stress relief temperature to reduce the width of the carburized zone
- applying a full heat treatment after welding to allow for chromium diffusion
- inserting diffusional barriers (e.g., Ni-layer) between base metal and filler metal

Conclusions

The comparison between experimental and numerical results has proven that the physical based model is able to predict the complex diffusion processes in a dissimilar weld. Moreover the ternary diffusion model provides more insight into the basic mechanisms involved and allows for the selection of appropriate means of controlling the microstructural development.

References

[1] C.D. Lundin, Welding Journal 61, 58s (1982).
[2] R.D. Nicholson, Metals Technology 11, 115 (1984).
[3] M. Witwer, "Untersuchungen an Mischschweißverbindungen warm-fester CrMoV-Stähle (GS-17CrMoV5 11 an X20CrMoV12 1)", Ph.D. Thesis, Graz University of Technology (1989).
[4] B. Buchmayr, H. Cerjak, J.S. Kirkaldy and M. Witwer, 2nd Int. Conf. Trends in Welding Research, Gatlingburg, Tenn. (1989).
[5] J.S. Kirkaldy, Can. Met. Quart., 8, 35 (1969).
[6] J.S. Kirkaldy, in Oxidation of Metals and Alloys, p. 10, ASM Ohio (1971).
[7] H. Wada, Met. Trans. 16A, 1479 (1985).
[8] B. Aronsson, Climax Molybdenum Special Publication, 77 (1973).
[9] K. Bungardt, E. Kunze and E. Horn, Archiv Eisenhüttenwesen 29, 193 (1958).
[10] W. Jellinghaus and H. Keller, Archiv Eisenhüttenwesen, 43, 319 (1972).
[11] C. Wert, Phys. Rev. 79, 601 (1950).
[12] A.W. Bowen and G.M. Leak, Met. Trans. 1, 2767 (1970).

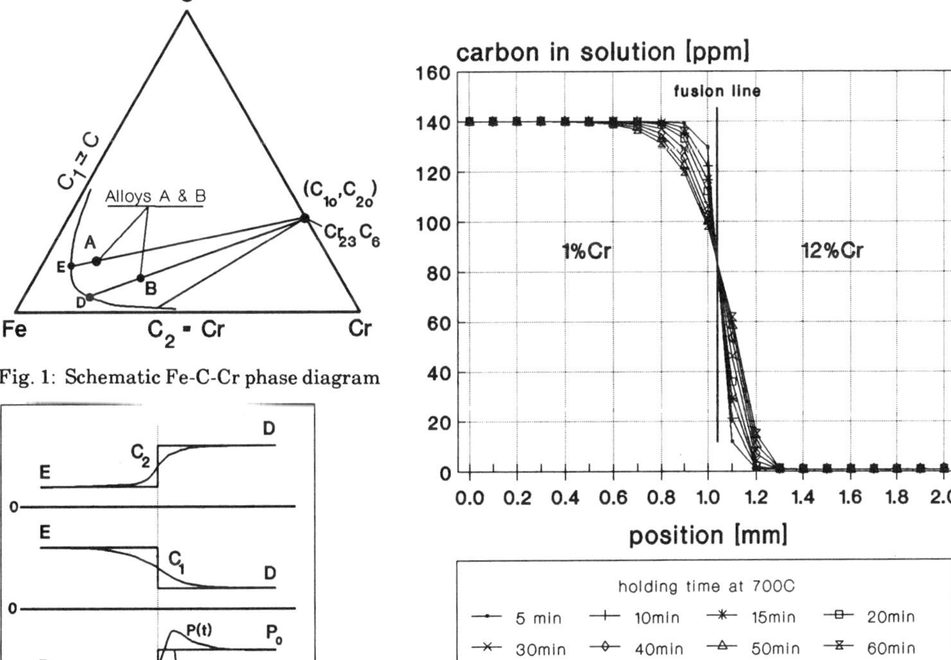

Fig. 1: Schematic Fe-C-Cr phase diagram

Fig. 2: Profiles of concentrations and volume fraction

Fig. 3a): Calc. profile of carbon dissolved in the ferrite matrix.

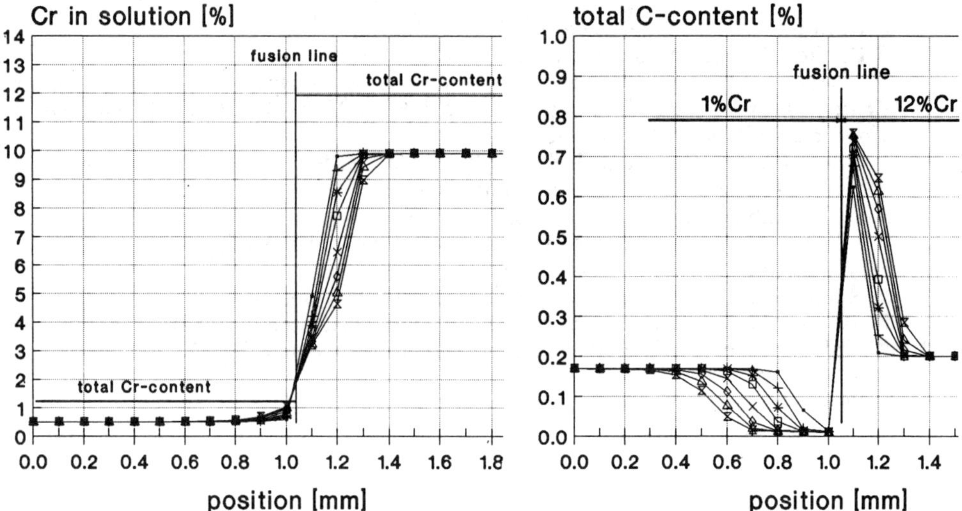

Fig. 3b): Calc. profile of chromium dissolved in the ferrite matrix

Fig. 4: Profile of total carbon concentration

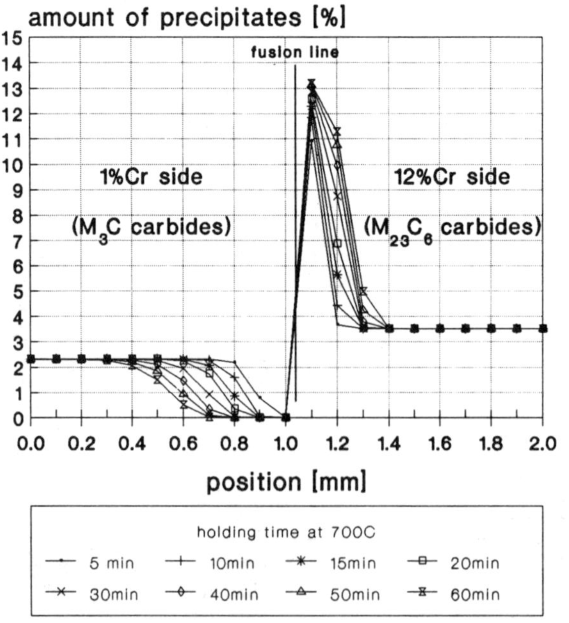

Fig. 5: Calc. amount of precipitates for the 1%Cr/12%Cr couple

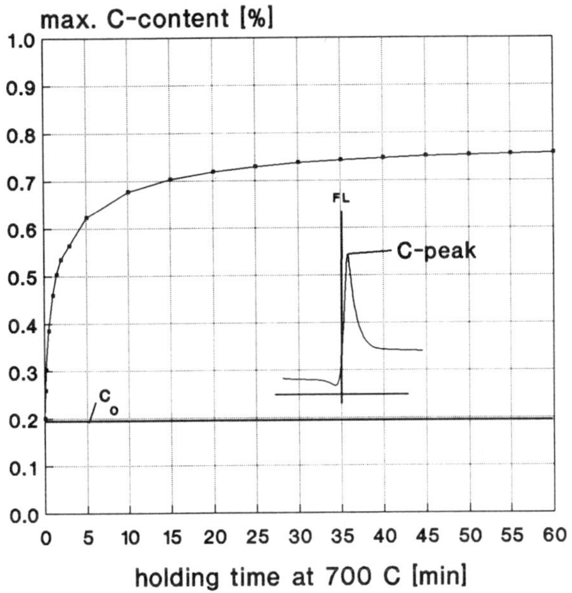

Fig. 6: Developing of maximum C-peak at the fusion line in the high Cr zone (calculated)

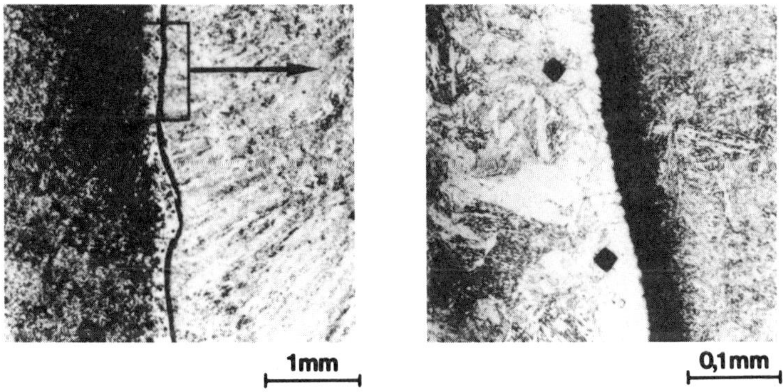

Fig. 7: Micrographs of the dissimilar weldment between 1%CrMoV and 12%CrMoV steel (3)

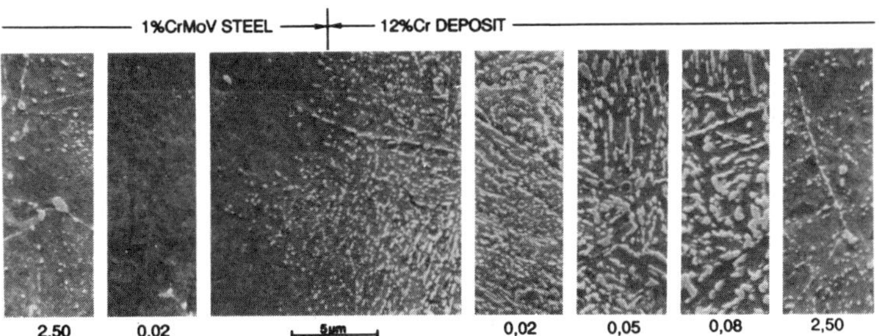

Fig. 8: SEM-micrographs of precipitation distribution as a function of the distance from the fusion line (3)

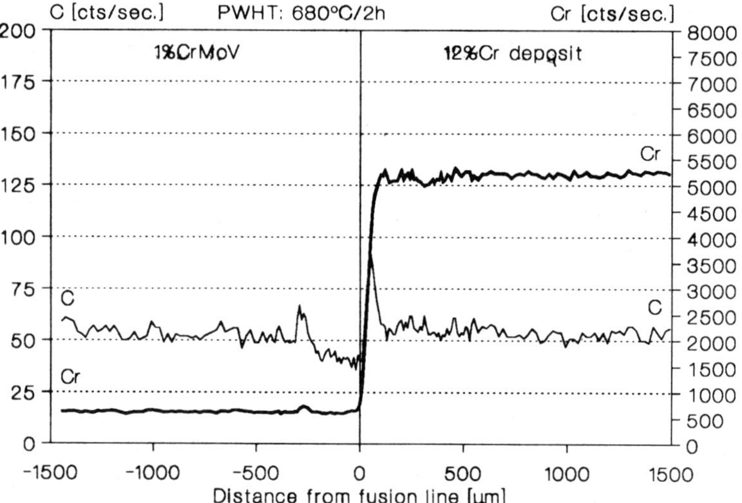

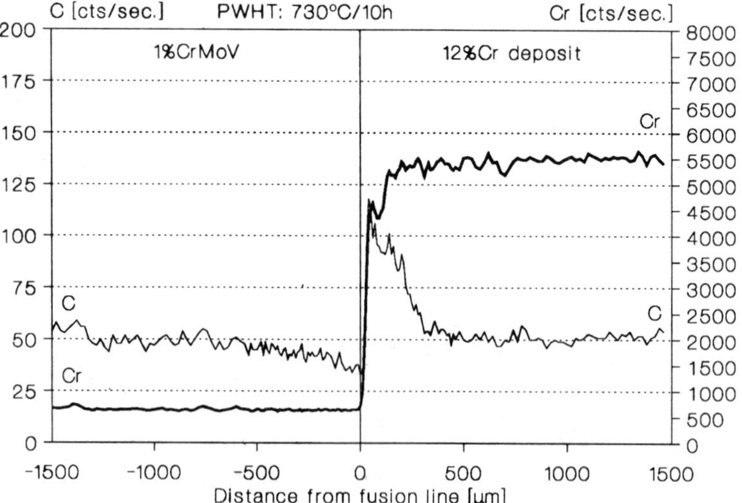

Fig. 9: Measured carbon and chromium profiles at the interface (3)
(a) PWHT: 680°C/2h b) PWHT: 730°C/10h

Decomposition of multicomponent austenite in spheroidal graphite cast iron

Dev Venugopalan
Materials Department, University of Wisconsin-Milwaukee, Milwaukee, Wisconsin 53201, U.S.A.

Abstract

In spheroidal graphite cast irons, diffusional decomposition of multicomponent austenite can occur along two paths available for the eutectoid reaction: one path resulting in ferrite and graphite as the reaction products while the other in a pearlite as the product. Under the prevalent conditions of commercial production of spheroidal graphite irons, both reactions occur leading to a mixed matrix microstructure consisting of both free ferrite and pearlite. The thermodynamics and kinetics of the two reactions are determined by the chemical composition of the austenite, grain size, number density of graphite spheroids in the iron and the cooling rate. In this paper, isothermal and continuous cooling transformations of multicomponent austenite are modelled based on alloy thermodynamics and on multicomponent diffusion theory. The quantitative results are compared with experimental results.

Introduction

Spheroidal graphite (SG) cast irons are multicomponent alloys based on the Fe-C-Si ternary system. This class of irons has evolved over the past four decades as an important engineering material with applications in automotive and heavy engineering industries. The properties of the iron depends on its chemical composition and microstructure, which consists of spheroids of graphite distributed in a metal matrix. In particular, the properties of the matrix determine the properties of SG iron. Thus it is of interest to the producer and user of SG iron to understand the development of the matrix microstructure to enable prediction and control of its microstructure and properties.

The matrix microstructure in SG iron is the product of austenite transformation as the iron is cooled either in the mold in which it is cast or in a heat treatment operation. Since the austenite in SG iron is nearly of the eutectoid composition, the transformation occurring as the austenite is cooled is predominantly the eutectoid reaction. In the Fe-C-Si ternary system, the eutectoid reaction has two branches; one branch leads to the decomposition of austenite to ferrite and graphite, which are the stable or equilibrium phases, and the other branch leads to decomposition of austenite to pearlite (a metastable mix of ferrite and cementite). The latter reaction (henceforth referred to as the pearlite reaction) is similar in many respects to the pearlite reaction in steels while the transformation leading to ferrite and graphite (henceforth referred to as the ferrite reaction) is generally not observed in steels.

The ferrite reaction can occur at temperatures below the A_1 temperature in the stable phase diagram (the ferrite A_1) and the pearlite reaction can occur below the metastable A_1 temperature (the pearlite A_1). In Fe-C-Si alloys, the ferrite A_1 temperature is higher than the pearlite A_1 temperature. The thermodynamic driving force for each reaction increases with increasing undercooling below the appropriate A_1 temperature. Both reactions occur by mechanisms of nucleation and growth.[1]

Ferrite Reaction

As austenite cools below the ferrite A_1 temperature, transformation of austenite to ferrite and graphite begins in the vicinity of graphite spheroids which act as sinks for the rejected carbon atoms (1). As the reaction proceeds further, the ferrite region develops as a spherical shell enveloping the graphite spheroid, resulting in the characteristic bull's eye structure as shown in Figure 1. Experimental studies of the isothermal kinetics of this reaction indicate the presence of an incubation period which is generally attributed to the nucleation and growth of ferrite to observable extent (usually 0.1%) (1,2). The growth of the ferrite shells is attended by a rejection of carbon atoms which diffuse through the ferrite shells to the existing graphite spheroids as well as into austenite (note that below A_1 temperatures, carbon solubility in austenite is greater than the eutectoid composition). The ferrite reaction proceeds without partitioning of the alloying elements (3).

The driving force for diffusion of carbon through the ferrite shells, ΔC_1, is the difference between the solubility of carbon in ferrite at the austenite/ferrite boundary and that at the graphite/ferrite boundary. If local equilibrium exists at the interfaces, the carbon concentration at the ferrite/austenite boundary is given by the metastable extension of the $\alpha/(\alpha+\gamma)$ boundary while the carbon concentration at the ferrite/graphite boundary is given by the solvus line of the stable iron-graphite phase diagram. Diffusion in austenite is driven by the concentration difference ΔC_2, between the metastable extension of $(\alpha+\gamma)/\gamma$ boundary in the phase diagram and the bulk austenite composition. It is possible to calculate the various equilibrium compositions and metastable extensions of the phase boundaries from solution thermodynamic models.

The effects of alloying elements on the kinetics of the ferrite reaction are due to their effect on the thermodynamic driving force (by altering the A_1 temperature) and on the diffusion coefficient of carbon. Elements which raise the A_1 temperature increase the driving force for the ferrite reaction at a given temperature. The transformation occurs at higher temperatures during cooling. Ferrite stabilizers such as silicon act in this way (4). Molybdenum, at levels less than 0.3%, also increases the transformation temperature but Mo retards carbon diffusivity (4). The net effect of Mo is observed to be increased transformation temperatures but longer transformation times. Austenite stabilizers such as Mn, Ni and Cu push the ferrite reaction to lower temperatures. Of these common alloying elements present in SG irons, Si increases the carbon diffusion coefficient at a given temperature while Mn and Mo decrease it. Ni does not affect C diffusivity significantly. Copper is attributed to retarding the diffusion of carbon near the graphite spheroids (due to formation of a Cu enriched zone around the spheroids) (5).

Since nucleation of ferrite occurs near the graphite spheroids, the number density of spheroids in the iron influences the ferrite reaction rate (6). It has been seen experimentally that increasing the spheroid density leads to an increase in the rate of ferrite formation (7).

Pearlite Reaction

As austenite cools below the A_1 temperature for the metastable eutectoid reaction, it becomes thermodynamically unstable with respect to pearlite also.

Further undercooling results in nucleation and growth of pearlite in the untransformed volumes of austenite. The pearlite reaction in SG iron is similar in to that in steel in the mechanisms of nucleation and growth. Nucleation of pearlite colonies occurs preferentially at austenite grain boundaries and occasionally at the ferrite/austenite boundaries if the ferrite reaction has preceded the pearlite reaction. In this paper, a kinetic model of pearlite reaction in steels developed by Kirkaldy and Venugopalan (8) is extended to describe the pearlite reaction in ductile iron.

Alloying elements affect the pearlite reaction through their effects on the thermodynamics and kinetics of the reaction. Austenite stabilizers such as Mn, Ni and Cu lower the transformation temperatures (4). These elements do not partition during the pearlite reaction and have a small effect on the diffusivity of carbon (9). Due to the lower transformation temperatures, the overall pearlite reaction rate is retarded and the transformation curves are shifted to longer times. Silicon also does not partition during the pearlite reaction but it increases the transformation temperature and enhances the diffusivity of carbon (9). These effects of silicon increase the reaction rate. Molybdenum, on the other hand, also increases the transformation temperature but it partitions during the pearlite reaction (4,9). The requirement of diffusion of Mo during the reaction leads to significantly slower reaction kinetics and the pearlite transformation curves are shifted to longer times.

Isothermal Transformation Rates

Ferrite Reaction

Isothermal ferrite reaction has been recently modeled by Venugopalan (6). The model assumes a uniform distribution of equisized graphite spheroids. Each graphite spheroid is surrounded by a volume of austenite which can be represented by an equivalent sphere of radius, R_e, given by Eqn. 1 (the shape of the austenite regions actually is a tetrakaidodecahedron, a polyhedron with 14 sides, which is enveloped by surfaces across which no diffusional flux occurs due to symmetry).

$$R_e = \left[\frac{3}{4\pi N_V}\right]^{(1/3)} \tag{1}$$

where N_V is the spheroid number density. The ferrite shell grows into austenite at a rate controlled by the flux of carbon atoms away from the austenite/ferrite interface. The incorporation of carbon atoms into graphite spheroid leads to growth of the spheroid at a rate dependent on the arrival rate of carbon atoms at the ferrite/graphite interface. With the assumption of local equilibrium condition at the interfaces, Eqns. 2 and 3 can be written for the mass conservation of carbon atoms at the austenite/ferrite interface and at the ferrite/graphite interface, respectively (6).

$$4\pi R_\alpha^2 \left\{ D^\alpha \left[\frac{dC^\alpha}{dr}\right]_{r=R_\alpha} + D^\gamma \left[\frac{dC^\gamma}{dr}\right]_{r=R_\alpha} \right\} = 4\pi R_\alpha^2 \left[C^{\gamma\alpha} - C^{\alpha\gamma}\right] \frac{dR_\alpha}{dt} \tag{2}$$

$$4\pi R_g^2\, D\left[\frac{dC^\alpha}{dr}\right]_{r=R_g} = 4\pi R_g^2 [1 - C^{\alpha g}]\, \frac{dR_g}{dt}\, A \tag{3}$$

where D^α is the diffusion coefficient of carbon in ferrite, D^γ is the diffusion coefficient of carbon in austenite, R_α is the outer radius of the ferrite shell, R_g is radius of the graphite spheroid, $C^{\alpha\gamma}$ is the carbon content of ferrite at the α/γ interface, $C^{\alpha g}$ is the ferrite carbon content at the ferrite/graphite interface, C^α is the carbon content of ferrite and C^γ is the carbon content of the austenite. In Eqn. 3, the factor A (which is about 3) is included in the right hand side to account for the difference in the densities of ferrite and graphite. The graphite and ferrite volume fractions can be related to their radii and the radius of the cell unit, R_e by Eqn. 4.

$$f_g = \frac{R_g^3}{R_e^3} \; ; \qquad f_\alpha = \frac{R_\alpha^3 - R_g^3}{R_e^3} \tag{4}$$

Equations 2 and 3 can be rearranged and integrated to obtain R_α and R_g as functions of time provided the concentration gradients in ferrite at the appropriate interfaces are known. If the time step used in the integration is sufficiently small, then the spherically symmetric concentration profile in the ferrite shell can be represented by the quasi-steady state profile as

$$C^\alpha = a + \frac{b}{r} \tag{5}$$

With the limits that at $r=R_g$, $C^\alpha = C^{\alpha g}$ and at $r=R_\alpha$, $C^\alpha = C^{\alpha\gamma}$, Eqn. 5 can be written as

$$C^\alpha = \frac{C^{\alpha\gamma}R_\alpha - C^{\alpha g}R_g}{R_\alpha - R_g} - \frac{\Delta C_1 R_\alpha R_g}{(R_\alpha - R_g)}\frac{1}{r} \tag{6}$$

where $\Delta C_1 = C^{\alpha\gamma} - C^{\alpha g}$. The concentration gradient in austenite at the α/γ boundary can be conveniently written (6), following Zener (10), as in Eqn. 7.

$$\left[\frac{dC^\gamma}{dr}\right] = (C^{\gamma\alpha} - C^{\gamma 0})\,\frac{1}{R_\alpha}\left[1 - \frac{R_\alpha}{R_e}\right] \tag{7}$$

Note that the expression in Eqn. 7 satisfies the requirement that the concentration gradient in austenite decrease as the reaction progresses due to the enrichment of the austenite shell. This is an *ad hoc* way of dealing with the diffusion problem in austenite but is justified because the diffusional flux in austenite is small (6) and this form of expression has been successfully used in literature to solve the problem of growth of spherical particles in mass conserved cells as in the present situation.

Combining the expressions for concentration gradients in ferrite and austenite with the growth equations (Eqns. 2 and 3) and incorporating the relations between volume fractions and radii given by Eqn. 4, results in Eqns. 8 and 9 for the rate of change of volume fractions of ferrite and graphite.

$$\frac{df_g}{dt} = \frac{3D^\alpha \Delta C_1 \Phi_1 \Phi_2}{A(1-C^{\alpha g})R_e^2(\Phi_1 - \Phi_2)} \tag{8}$$

$$\frac{d(f_\alpha + f_g)}{dt} = \frac{3D^\alpha \Delta C_1 \Phi_1}{\Delta C_3 R_e^2}\left[\frac{\Phi_2}{\Phi_1 - \Phi_2} + \frac{D^\gamma}{D^\alpha}(1 - \Phi_1)\right] \tag{9}$$

where $\Delta C_1 = C^{\alpha\gamma} - C^{\gamma g}$; $\Delta C_2 = C^{\gamma\alpha} - C^{\gamma o}$ and $\Delta C_3 = C^{\gamma\alpha} - C^{\alpha\gamma}$. Φ_1 and Φ_2 are functions of the volume fractions given by Eqn. 10.

$$\Phi_1 = [f_\alpha + f_g]^{1/3} \qquad \Phi_2 = f_g^{1/3} \tag{10}$$

The isothermal rates of formation of ferrite and graphite can be obtained by integrating Eqns. 8 and 9 simultaneously by numerical methods. The initial conditions for integration are that at time $t=0$, $f_g=f_{go}$ which is the initial graphite volume fraction, and at $t=t_i$, $f_\alpha=f_{\alpha i}$, where t_i is the "incubation" period for transformation to occur so that the ferrite volume fraction is $f_{\alpha i}$. The initial time is taken as that at which the ferrite reaction has proceeded to the extent of having an observable ferrite volume percent, which is usually 0.1 or 1%. The time for transformation to this percentage is usually taken as a multiple of the time required for nucleation alone. The latter is expressed as an inverse function of the nucleation rate from classical nucleation theory. It is given formally as (6)

$$t_i = \frac{constant}{D^\gamma \Delta T^n} \tag{11}$$

where D^γ is the carbon diffusion coefficient in austenite and ΔT is the degree of undercooling below the stable eutectoid temperature. From experimental data (1) on isothermal transformation in a SG iron with 1% Ni and 2.5% Si, n was obtained as 4 and the constant as 2.7 using the following expression for austenite carbon diffusion coefficient

$$D^\gamma = 0.3\ e^{-18,000/T} \tag{12}$$

It is to be noted that the transformation start time is a function of the composition of the iron and possibly its thermal history. The values of n and of the constant in Eqn. 10 have to be determined experimentally.

With these initial conditions, the growth rate equations were integrated for the conditions corresponding to the iron used in ref. 1. The results of the calculations are compared with the available experimental data in Fig. 2. The A_1 temperature and the relevant phase boundary compositions used in the calculations were determined from a calculation of the phase diagrams (11) using the model of Hillert and Staffansson (12) and thermodynamic data compiled by Uhrenius (13). It was assumed that paraequilibrium conditions prevail during the ferrite reaction (note that the reaction occurs without diffusion of alloying elements). The paraequilibrium condition is well described mathematically by Gilmour et al (14).

The agreement between the calculated volume fractions and experimentally measured values is excellent indicating the validity of the growth equations. It is noteworthy that the only parameter not derived from independently is the incubation time. This is due to the difficulty of modelling solid state nucleation process to obtain quantitatively meaningful results. The present growth model also predicts that the rate of formation of ferrite increases with increasing spheroid density. Note that R_e is inversely related to N_v via Eqn. 1 and that the growth rates are inversely related to R_e via Eqns. 8 and 9. Increase in ferrite formation rate with increasing spheroid number density is qualitatively verified by experiment.

The influence of alloying elements on the kinetics of the ferrite reaction is easily dealt with in this model. Since the phase boundary compositions and A_1 temperatures are determined by actually calculating the multicomponent phase diagram, the thermodynamic effects of alloying elements on the ferrite reaction are accounted for without additional modifications to the model. Alloying elements also affect the diffusion coefficient of carbon and such data are not available for multicomponent austenite and ferrite. However, this effect can be incorporated in a semi-empirical way by calibrating the model to experimental data on the ferrite reaction. The effective diffusion coefficient of carbon can be expressed as (assuming that the activation energy for diffusion does not change with alloy content)

$$D = D_0[1 + \Sigma\, a_i C_i]\, e^{-Q/RT} \tag{13}$$

where C_i is the amount of the i-th alloying element in austenite and a_i is the extent to which this element influences the diffusivity of carbon.[7] This procedure was resorted to in the calculation of cooling transformation diagrams in a later section of this paper.

Pearlite Reaction

The isothermal kinetics of transformation of austenite to pearlite in steels has been described by Kirkaldy and Venugopalan (8) by a semi-empirical model which incorporates the results from solution thermodynamics on driving forces for the reaction and from a formal treatment of the pearlite growth problem. It is assumed that pearlite colonies nucleate in untransformed volumes of austenite by a point nucleation mechanism and grow as spherical colonies. This model led to Eqn. 14 for the growth rate of pearlite in which ΔT is the undercooling below the pearlite A_1 temperature, G is the ASTM austenite grain size number, m is given as $2(1-f_p)/3$ and q as $2f_p/3$.

$$\frac{df_p}{dt} = 2^{(G-1)/2} K D \Delta T^3 f_p^m (1-f_p)^q \tag{14}$$

In Eqn. 14, D is an effective diffusion coefficient for the pearlite reaction. The effective diffusion coefficient was derived based on an analysis of multicomponent diffusion and the nature of pearlite reaction. Among the alloying elements which are normally added to SG iron, Mo is the only one which partitions during the pearlite reaction with diffusion Mo occurring along the phase boundaries. As a result, the diffusion resistance in the reaction is obtained by addition of series resistances due to carbon diffusion and due to Mo diffusion. The effective diffusion coefficient then can be written as

$$\frac{1}{D} = \frac{1}{D_C} + \frac{b_{Mo} C_{Mo}}{D^b_{Mo}} \tag{15}$$

where D^b_{Mo} is the boundary diffusion coefficient of Mo, D_C is the carbon diffusion coefficient, b_{Mo} is a coefficient to be determined empirically and C_{Mo} is the amount of Mo in the austenite. The constant K in Eqn. 14 is again a factor which accounts for the effect of other alloying elements on the effective diffusion coefficient and it can be expressed as a first order series in the concentrations of the alloying elements. The detailed expressions for isothermal pearlite reaction are presented in ref. 8. As with the ferrite reaction, the A_1 temperature was derived from thermodynamic models.

The success of the model in accurately predicting the kinetics of the pearlite reaction has been shown by Kirkaldy and Venugopalan (8) who compared the predictions of the model with experimental transformation curves from a number of different data sets. Since the pearlite reaction in SG irons is similar to that in steels, the model developed for steels is expected to apply to SG irons with the same degree of success.

Overall Isothermal Transformation

The transformation equations for ferrite and pearlite can be integrated to obtain the overall austenite transformation diagram at any temperature. At temperatures above the A_1 temperature for pearlite, the reaction products are only ferrite and graphite. At temperatures below the pearlite A_1 temperature, a mix of products can form. If the starting time for the pearlite reaction is less than the starting time for the ferrite reaction, then it is assumed that the entire matrix transforms to pearlite. If pearlite start time is greater than ferrite finish time, then the entire matrix is made up of ferrite only. In the other cases, both ferrite and pearlite growth equations are integrated simultaneously and when the total volume fraction transformed reaches unity, the corresponding time is taken as the finish time for the transformation. The individual volume fractions of pearlite and ferrite are obtained from the instantaneous values of the f_p and f_α respectively.

Continuous Cooling Transformation

Continuous cooling transformation (CCT) diagrams can be obtained from isothermal transformation diagrams by applying the additivity rule which was originally formulated by Avrami (15) for general chemical reactions and later applied specifically for transformations in steels by Kirkaldy and Sharma (16). The additivity rule has been shown to be applicable to transformations occurring by nucleation and growth processes. The isothermal transformation diagrams for ferrite and pearlite formation can be expressed by a function $\tau(X,T)$ where τ is the time elapsed in producing a volume fraction X of the product at a temperature of T. In most practical situations, the transformation occurs over a range of temperatures along a cooling path.

The additivity rule states that in the time Δt which the iron spends at T, a reaction that results in X volume fraction of a product phase will occur to the extent of $\Delta\varphi = \Delta t/\tau(X,T)$. The total extent of this transformation occurring after cooling through a temperature range is given by the summation

$$\varphi = \Sigma\, \Delta\varphi = \Sigma \frac{\Delta t}{\tau(X,T)} = \Sigma \frac{\Delta T}{\theta\tau(X,T)} \tag{15}$$

where θ is the cooling rate at T. When φ reaches unity, the reaction resulting in X volume fraction of the product is considered to have been completed. For example, consider the transformation of austenite to form 1% ferrite in the microstructure. From the isothermal transformation diagrams, the value of $\tau(0.01,T)$ can be obtained. To determine the time and temperature at which 1% ferrite is formed for a given cooling path, the summation in Eqn. 11 is carried out for each step along the cooling path until the value of φ becomes unity. The time and temperature at which φ reaches unity corresponds to the time and temperature at which 1% ferrite is formed for the chosen cooling path. If φ does not reach unity, then the cooling rate is too rapid to allow formation of 1% ferrite.

To construct the CCT diagram for the iron, this procedure must be followed for a set of cooling paths and for all volume fractions transformed. In the case of mixed products, such as ferrite and pearlite, the calculations are conducted simultaneously for both products and the overall transformation curves are obtained accordingly.

Results and Discussion

The equations derived above were integrated using a personal computer. The compositions and equilibrium temperatures were obtained by solving the appropriate thermodynamic equations. The calculated isothermal kinetics of ferrite formation are compared with the experimental data in Fig. 2 which shows excellent agreement between the calculated and measured results. Direct verification of the model for pearlite formation in SG irons is not possible due to lack of reported experimental data. However, as stated above, the isothermal model for pearlite growth was compared well with data for steel.

While isothermal transformation curves are not available in the literature, a compilation of CCT diagrams for constant cooling rates has been presented for unalloyed, Mo and Ni-Mo ductile irons (17). As a test of the model presented here, CCT diagrams were calculated for the irons for which data are available. The kinetic coefficients (a_i's and b_{Mo}) for the transformations were derived by calibrating the model with the CCT start curves for the set of 20 ductile irons and optimizing the coefficients. The results for an unalloyed iron, a 0.5% Mo iron and a 1.2% Ni-0.5% Mo iron are given in Figs. 3 to 5 as examples of the good agreement between the calculations and the experimental results. In Figs. 3 to 5, the start and finish times of the ferrite and pearlite reactions for the indicated irons are shown. The calculated ferrite percents in the microstructure are plotted against the values estimated from the micrographs in Fig. 6 (the pearlite percent is given as 100-%Ferrite). Again the agreement between calculations and experiment is excellent.

Discussion

The present model is a very general model and easily incorporates the effects of alloying elements and spheroid density. The accuracy of the predictions can be improved by calibrating the model to a larger set of data from a given plant. From the comparisons made between the model predictions and the limited set of experimental data, it can be concluded that the model is capable of accurately predicting the microstructure in SG irons resulting from the eutectoid reactions. It is salutary to the model that it is based soundly on thermodynamics of alloy systems and on applied multicomponent diffusion theory and is able to make accurate predictions with minimal requirement for empirical coefficients. Thus the model is able to take into account the effects of alloying elements in the most reasonable way.

Acknowledgement

The author has, over the years, benefitted greatly from numerous stimulating discussions with Professor Jack Kirkaldy on the application of thermodynamics and diffusion theory to the study of phase transformations.

References

1. B.F. Brown and M.F. Hawkes, AFS Transactions, vol. 59, p. 181 (1951).
2. E.I. Eckel, AFS Transactions, vol. 66, pp. 151-159 (1958).
3. W.C. Johnson and B.V. Kovacs, Metall. Trans. A, vol. 9A, p. 219 (1978).
4. K. Rohrig, H.G. Gerlach and O. Nickel, Legiertes Gusseisen, Volume 2, Giesserei-Verlag GMBH, Dusseldorf (1974).
5. M.J. Lalich and C.R. Loper, Jr., AFS Transactions, vol. 81, p. 217 (1973).
6. D. Venugopalan, Metall. Trans., vol. 21A, p. 913 (1990).
7. M.J. Lalich, and C.R. Loper, Jr., AFS Transactions, vol. 81, p. 238 (1973).
8. J.S. Kirkaldy and D. Venugopalan, in Phase Transformations in Ferrous Alloys, A.R. Marder and J.I. Goldstein, Eds., TMS-AIME, Warrendale, PA, p. 125 (1984).
9. M.P. Puls and J.S. Kirkaldy, Metall. Trans., vol. 3, p.2777 (1972).
10. C. Zener, J. Appl. Phys., Vol. 20, p. 950 (1949).
11. D. Venugopalan, unpublished research (1989).

12. M. Hillert and L.I. Staffansson, Acta Chem. Scan., vol. 24, p. 3618 (1970).
13. B. Uhrenius, in Hardenability Concepts with Applications to Steel, D.V. Doane and J.S. Kirkaldy, Eds., TMS-AIME, Warrendale, PA, p. 28 (1977).
14. J.B. Gilmour, G.R. Purdy and J.S. Kirkaldy, Metall. Trans., vol. 3, p. 1455 (1972).
15. M. Avrami, J. Chem. Phys., vol. 7, p. 1103 (1939).
16. J.S. Kirkaldy and R.C. Sharma, Scripta Met., vol. 16, p. 1193 (1982).
17. Continuous Cooling Transformation Diagrams for Ductile Irons, Climax Molybdenum Co. (1982).

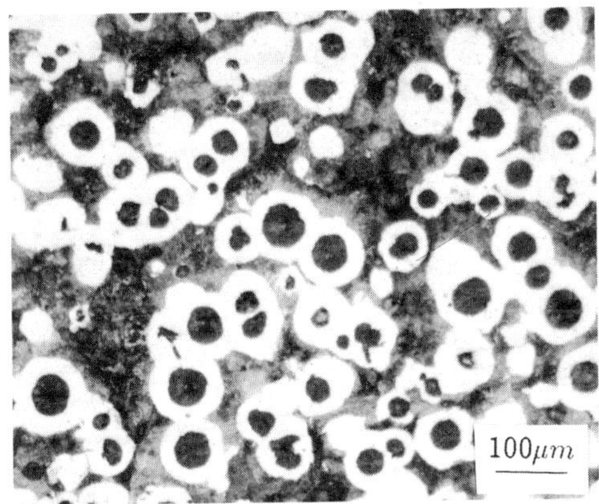

FIG. 1. Micrograph of SG iron showing ferrite shells around graphite spheroids. The rest of the matrix is pearlite (dark).

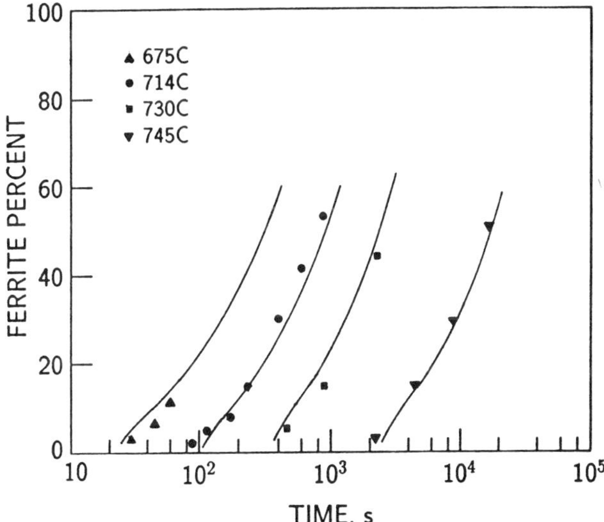

FIG. 2. Comparison of calculated ferrite volume fractions in isothermal transformations at indicated temperatures with experimental data from ref. 1 (6).

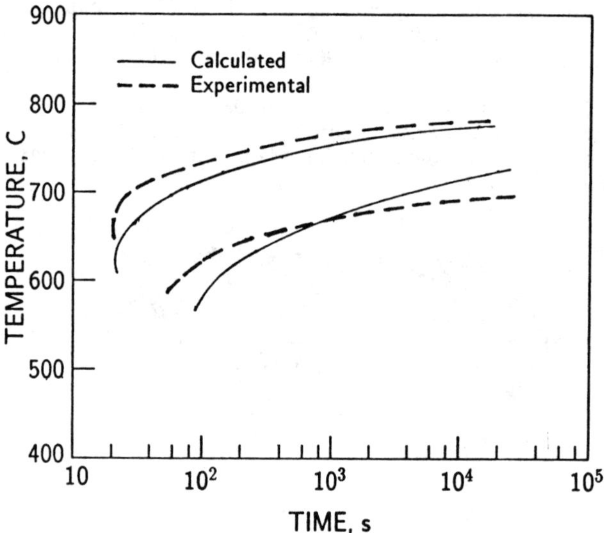

FIG. 3. Comparison between calculated and experimental (17) CCT diagram for SG iron with 3.37 C, 2.62 Si and 0.31 Mn.

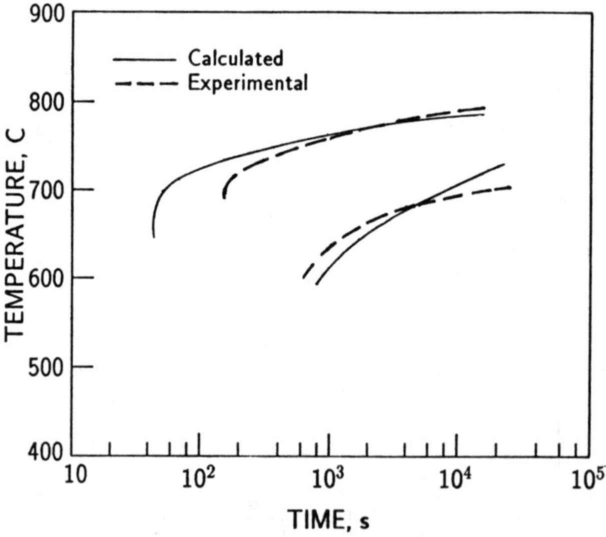

FIG. 4. Comparison between calculated and experimental (17) CCT diagram for SG iron with 3.32 C, 2.58 Si, 0.31 Mn and 0.49 Mo.

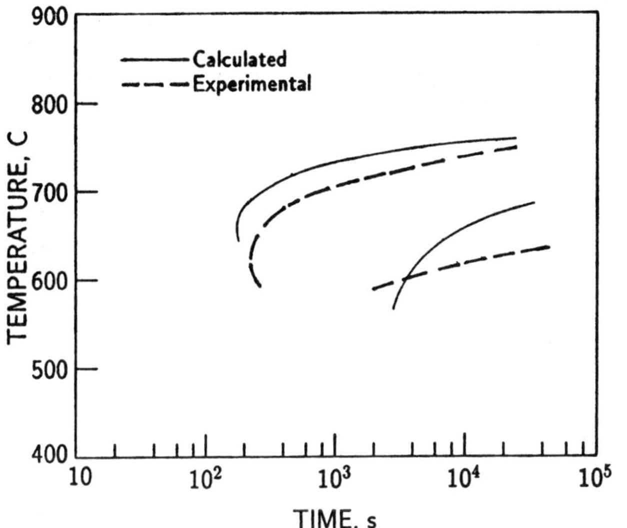

FIG. 5. Comparison between calculated and experimental (17) CCT diagram for SG iron with 3.36 C, 2.46 Si, 0.32 Mn, 1.17 Ni and 0.49 Mo.

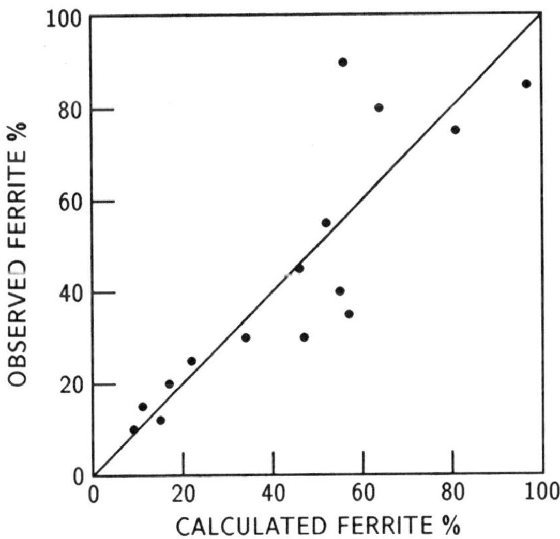

FIG. 6. Comparison between calculated and oberved (17) ferrite contents of fifteen SG irons of various compositions.

Multicomponent diffusion theory applied to carbonitride precipitate growth in pentenary microalloyed steels

Heilong Zou and J.S. Kirkaldy
Department of Materials Science and Engineering, McMaster University, Hamilton, Ontario, Canada, L8S 4M1

Abstract

The isothermal precipitation of spherical carbonitride particles in Fe has been modelled taking account of multicomponent diffusion effects in the fcc matrix and a solubility modified constitution including the effect of capillarity. Quantitative agreement between observations and the predictions for average particle size in the pre-coarsening stage is demonstrated. A sensitivity analysis shows that the multicomponent effects are significant despite the low concentrations of alloy additions. From this and other similar results it is concluded that multicomponent diffusion effects should be accounted for in all relevant first order phase transformations.

Introduction

Purdy et al. (1) offered the first quantitative application of Onsager's multicomponent diffusion theory (2,3) to the prediction of a phase transformation rate. This pertained to the isothermal transformation of the γ (fcc) to the α (bcc) in a dilute carbon-manganese alloy with iron (i.e., steel). It was demonstrated that the addition of manganese reduces the transformation rate of the first order reaction for two reasons (the so-called hardenability effect). In the first instance Mn is a γ (austenite) stabilizer (as is carbon) and therefore for a given temperature the supersaturation is reduced by the manganese addition. Secondly, Mn and C are attractive in the Fe lattice (C tends to diffuse up a Mn gradient), so in the present reaction where Mn and C are both rejected by the product α-phase, fast-diffusing C is attracted against the flow and thus a further resistance to transformation is made evident. These effects are represented analytically within the theory presented by Purdy et al. (1) and confirmed by experiment.

The two effects are illustrated comparatively (in Fig. 1) in terms of a normalized rate parameter β for a high supersaturation alloy of 0.2 wt% C and 1.5 wt% Mn as a function of temperature. As can be seen the constitutional retardation effect is about a factor of five while the diffusional effect is a factor of two to three. Despite the diluteness of the solution the diffusional cross-effect is highly significant.

In the present investigation we are concerned with a higher temperature reaction in γ-iron or austenite in which the alloying elements (typically very dilute and numbering four) are much more reactive. This has the effect that the precipitates are very much smaller so capillarity has a significant constitutional influence on the kinetics.

The latter considerations, like the strong retardation or quench hardening effect of Mn (4), have technological significance, for an ideal combination of strength and toughness in steels can be obtained by the hardening effects of a fine dispersion of precipitates, typically a dispersion of very fine and stable carbides, nitrides and carbonitrides precipitating from Nb, Ti and V microalloyed steels. The predictive optimization of microalloying element additions demands taking into account accurate solubility data as well as the kinetic evolution of microstructure and the various sequences of precipitation reaction in which the microalloying elements are involved. Accordingly we undertake the development of a theoretical model to predict the kinetics of carbonitride precipitation based on rigorous equilibrium thermo dynamics and a comparison with experiment. In summary the objective is to computer simulate the growth of multicomponent carbonitride precipitates in austenite alloyed with titanium and niobium and to compare the theoretical predictions with experiment. The experiments consist of the micrography and the growth kinetics of $(Ti_xNb_{1-x})(C_yN_{1-y})$ precipitates using TEM and STEM (5).

The Austenite - Carbonitride Equilibrium

The carbonitride may be described by the chemical formula $(Ti_xNb_{1-x})(C_yN_{1-y})$ (assuming perfect stoichiometry), where $0 \le x \le 1$ and $0 \le y \le 1$. With the inclusion of capillarity for spherical particles, the following four chemical potential equalities in terms of the composition of the two phases modelled according to a sub-regular solution (5,6) define the austenite-carbonitride equilibrium

$$RT\ln[X_C X_{Ti}] = \Delta G^o_{TiC} + RT\ln[xy] + \Delta G(1-x)(1-y) + \Omega(1-y)^2 + \frac{4\sigma V_c}{r} \tag{1}$$

$$RT\ln[X_C X_{Nb}] = \Delta G^o_{NbC} + RT\ln[(1-x)y] - \Delta G(x)(1-y) + \Omega(1-y)^2 + \frac{4\sigma V_c}{r} \tag{2}$$

$$RT\ln[X_N X_{Ti}] = \Delta G^o_{TiN} + RT\ln[x(1-y)] - \Delta G(1-x)(y) + \Omega y^2 + \frac{4\sigma V_c}{r} \tag{3}$$

$$RT\ln[X_N X_{Nb}] = \Delta G^o_{NbN} + RT\ln[(1-x)(1-y)] + \Delta G(x)(y) + \Omega y^2 + \frac{4\sigma V_c}{r} \tag{4}$$

where ΔG can be calculated from

$$\Delta G = \Delta G^o_{NbN} + \Delta G^o_{TiC} - \Delta G^o_{NbC} - \Delta G^o_{TiN} \tag{5}$$

The ΔG°_{MX} refer to the free energy of formation of compound MX from austenite with the reference state at infinite dilution, σ is the specific interfacial free energy of the precipitate/matrix interface, V_c is the molar volume of carbonitride, r is the radius of the particle, Ω is the empirical regular solution parameter and X_i is the mole fraction of the ith element in austenite at the precipitate interface at equilibrium with spherical precipitate. It is to be noted that only three of the above equations are independent.

The four mass balance equations can be written as

$$X^o_C = X_C(1-Z) + yZ/2 \tag{6}$$

$$X^o_N = X_N(1-Z) + (1-y)Z/2 \quad (7)$$

$$X^o_{Ti} = X_{Ti}(1-Z) + xZ/2 \quad (8)$$

$$X^o_{Nb} = X_{Nb}(1-Z) + (1-x)Z/2 \quad (9)$$

where the X°_i refer to the overall (or initial) composition of steel and Z refers to the mole fraction of the carbonitride. Equations (1) through (9) can be solved numerically to determine the austenite/carbonitride equilibrium.

The Generalized Fick Diffusion Equations

Onsager's form of the thermodynamics of irreversible processes (2) establishes a generalization wherein for an n-component system (n is the designated solvent) the flux is

$$J_i = -\sum_{j=1}^{n-1} D_{ij} \nabla C_j \quad (10)$$

where the D_{ij} are diffusion coefficients and the C_j are concentrations in moles/unit volume. The off-diagonal diffusion coefficients D_{ij} $(i \neq j)$ specify the diffusivity for flow of one component on the concentration gradient of the other component. For a dilute austenite system in which the changes in concentration are small, Eq. (4) can be applied to fluxes measured with respect to a fixed matrix lattice, i.e., lattice parameter changes and the shift of the Matano interface are ignored.

For dilute Fe-M-X ternary systems (X_M, $X_X << 1$), it has been shown (7) that

$$D_{XM} \simeq \varepsilon_{XM} X_X D_{XX} \quad (11)$$

$$D_{MX} \simeq \varepsilon_{MX} X_X D_{MM} \quad (12)$$

where the $\varepsilon_{ij} = \varepsilon_{ji}$ are the constant Wagner interaction parameters defined by the linear Taylor expansion of ln γ_i (γ_i^o is the Henry's law coefficient). The same approximation can be extended to the dilute Fe-Ti-Nb-C-N pentenary system, viz.,

$$D_{ij} \simeq \varepsilon_{ij} X_i D_{ii} \quad (13)$$

In this system, the fact that $D_{XX} >> D_{MM}$ assures that the effect of D_{XM} on the interstitial fluxes will be large and that of D_{MX} on the corresponding substitutional fluxes can be neglected. D_{CN}, D_{NC}, D_{NbTi} and D_{TiNb} can also be neglected since the cross interactions between Ti and Nb, and C and N are moderated by the near equality of on-diagonal D's. Thus the fluxes can be approximately represented as:

$$J_C = -D_{CC}(\nabla C_C + \varepsilon_{CTi} X_C \nabla C_{Ti} + \varepsilon_{CNb} X_C \nabla C_{Nb}) \quad (14)$$

$$J_N = -D_{NN}(\nabla C_N + \varepsilon_{NTi} X_N \nabla C_{Ti} + \varepsilon_{NNb} X_N \nabla C_{Nb}) \quad (15)$$

$$J_{Ti} = -D_{TiTi} \nabla C_{Ti} \quad (16)$$

$$J_{Nb} = -D_{NbNb} \nabla C_{Nb} \tag{17}$$

The Computational Procedure

In modelling the precipitation process we undertake the simultaneous solution of a set of diffusion, equilibrium and mass balance equations involving time-dependent boundary conditions. These equations are non-linear and transcendental so iterative methods have to be used. In summary we assume

1) The rate of precipitation process is controlled by volume diffusion of the Ti, Nb, C and N atoms in austenite, according to Eqs. (14-17).

2) Local equilibrium including the effect of capillarity holds at the interface, according to Eqs. (1-4).

3) The particles are spherical and the interfacial energy, σ, is a constant.

4) There is no diffusion within the particle.

We further assume that all the nuclei form immediately after the up-quenching with the same growth rate (site saturation) and are uniformly distributed in a Wigner-Seitz array (Fig. 2) such that the fluxes vanish at the cell limits (R_o). The problem is thus simplified so one needs to deal with only one cell.

Following the method of Voice and Faulkner (8) for carbide dissolution a computation process is constructed subject to the above assumptions. In one cell a carbonitride nucleus is considered to exist in a matrix of initially uniform composition (Fig. 3a). The matrix is theoretically divided into shells of equal thickness radiating from the center of the particle (typically of 2 nm thickness). Initial concentrations of solute's M and X are then assigned to each shell, irrespective of whether it is in a particle or matrix region (Fig. 3b). The one-dimensional form of Eqs. (14-17) are modified to the three-dimensional reality by weighting diffusion from one shell (i + 1) to i across the spherical interface of local area A_i. Accordingly the solute transferred through the interface A_i can be represented by

$$\text{ith solute transferred}/\Delta t = J_i A_i = -A_i \sum_{k=1}^{4} D_{ik} \frac{dC_i}{dx} \tag{18}$$

where dC_i/dx is the concentration difference/shell width, and Δt the time interval. This equation may be approximated by replacing C_i by X_i/V_m whence

$$\text{ith solute transferred}/\Delta t = -\frac{A_i D_i}{V_m \Delta x} [(X^i_{i+1} - X^i_i) + \Sigma\, \varepsilon_{ij} X^i_i (X^j_{i+1} - X^j_i)]\,, \quad (j \neq i) \tag{19}$$

where X^j_i is mole fraction of jth solute in the ith shell, Δx is the thickness of the shell, and V_m is the molar volume of the austenite matrix. The transferred solutes are subtracted uniformly from the (i + 1)th shell and added to the ith shell, whence its atoms are transferred to the (i–1)th shell in a similar way, and finally the diffused solutes are added to the particle surface. When the ultimate shell is reached at R_o, this is given the same atom density as the penultimate shell according to the condition that there can be no net flux into or out of each cell. Diffusion to the interface is carried out for an interval, Δt, determined by the time necessary to sustain the stability of the computation. This process maintains a 'smooth' concentration profile and is repeated until an iterative equilibrium is reached (as in Fig. 3c).

The stoichiometry restrictions of (Ti_xNb_{1-x}) $(C_y\ N_{1-y})$ require that the diffusional fluxes at the precipitate/matrix interface satisfy the equations

$$(1-x)J_{Ti} = x\,J_{Nb} \tag{20}$$

$$(1-y)J_c = y\,J_N \tag{21}$$

$$J_{Ti} + J_{Nb} = J_C + J_N \tag{22}$$

Eqs. (1-3) and (20-22) are six equations in six unknowns (x, y and four X_i) which must be solved for numerically consistent with Eqs. (14-17). Applying the conservation of M solutes to the whole particle the velocity of growth of a spherical precipitate is:

$$v = dr/dt = V_C(J_{Ti} + J_{Nb}) \tag{23}$$

Eqs. (1-3) and (20-22) are solved for each time increment Δt and at the same time the fluxes are calculated via Eqs. (14-17). Following the solution of the equilibrium and diffusion equations the position of the interface is then updated via Eq. (23) by adding $v\Delta t$ to the previous value, and hereby the variation of precipitate size with time is obtained. Simultaneously the internal composition parameters, x and y, at a certain position r in the particle are obtained yielding the composition profile in the particle (assuming negligible diffusion).

The required solubility data expressed in the form of free energies of formation of TiC, NbC, TiN and NbN in austenite are given in Table 1. The diffusion coefficients and the interaction parameters of C, N, Ti and Nb in austenite are given in Tables 2 and 3.

TABLE 1
Free Energies of Formation of TiC, NbC, TiN and NbN in Austenite

Compound	$\Delta G°_{MX}$ (J/mole)	Reference
NbC	$\Delta G°_{NbC} = -176335 + 8\,6\,T$	†
TiC	$\Delta G°_{TiC} = -\ 166923 + 14.64\,T$	Sharaiwa 11
NbN	$\Delta G°_{NbN} = -\ 147010 - 247.7\,T + 30.3\,T$	Bala 12
TiN	$\Delta G°_{TiN} = -221560 - 238.2\,T + 30.3\,T$	Bala 12

† Estimated using the data of Johansen and Smith in Refs. 9 and 10.

TABLE 2
Diffusion Coefficients of C, N, Ti and Nb in Austenite

Solute	Do (mm^2)	Q (kJ/mole)	Reference
C	67	157	Smith 13
N	91	169	Darken 14
Ti	15	250	Moll 15
Nb	75	264	Kurokawa 16

TABLE 3
Wagner Interaction Parameters Adopted in the Calculation

Wagner Interaction Parameter (17)	Reference
$\varepsilon_{CNb} = -66257/T$	Greenband 18
$\varepsilon_{CTi} = -79150/T$	Bala 19
$\varepsilon_{NNb} = -406240/T + 206$	Bala 20
$\varepsilon_{NTi} = -705{,}000/T + 385$	Bala 20

Results of the Computations

We have chosen to examine the multicomponent diffusion model in detail for a single Fe based alloy from our own experimental data set, which is to be reported elsewhere (5). This was chosen for its expected large cross effect, and has the composition in wt%, Ti – 0.13, Nb – 0.070, C – 0.075, N – 0.0016.

The yet unspecified parameters used in the computations were obtained or estimated in the following ways. The value $\Omega = -4260$J/mole for C-N mixing in the Ti-C-N system is adopted, following Roberts and Sandberg (21). The molar volume of the matrix $V_m = 7.31 \times 10^{-6}$ m^3/mole was calculated from the lattice parameter of austenite a = 0.365 nm, and the molar volume of carbonitride, V_c, was calculated from the carbonitride lattice parameter which is a function of the particle composition. In the quaternary Ti-Nb-C-N system, the lattice parameter of mixed interstitial phase $(Ti_xNb_{1-x})(C_yN_{1-y})$ can be fitted to

$$a_{(TiNb)(CN)} = xya_{TiC} + (1-x)ya_{NbC} + x(1-y)a_{TiN} + (1-x)(1-y)a_{NbN}$$

where the a_{MX} are the lattice parameters of MX. Here x and y are calculated from the equilibrium model on the appropriate isotherm. The lattice parameters for the binary carbides and nitrides are given in Table 4. The interfacial energy was taken to be 0.3 J/m^2 (with reference to data on austenite/$Cr_{23}C_6$ (25)). A shell spacing of 2 nm was used throughout the calculations, and a different time interval of 10^{-6} to 10^{-4} s was used depending on the stability of the computation. The half spacing, R_o, between particles was estimated to be 40 nm at 1000°C and 50 nm at 1100°C according to both the calculations and micrographs obtained in the broader study (5). The critical radius of precipitate, r_c, was calculated from Eqs. (1-3) by substituting the initial compositions X°_i. We then set $r = 1.01\, r_c$ to start the computation.

Fig. 4 shows the time evolution of the composition profiles across the matrix γ. As a sensitivity test we have repeated the calculation setting the $\varepsilon_{ij} = 0$ yielding the dashed line of Fig. 4b for the interstitials. For $\varepsilon = 0$, the C and N profiles are almost flat. However, for $\varepsilon \neq 0$ we note the expected depression of the profiles of C and N in the matrix matching the Ti and Nb profiles near the interface, and this local redistribution of the interstitials is directly due to the cross-interaction terms in Eqs. (14) and (15). The Ti and Nb profiles (Fig. 4(a)) are essentially independent of the cross-interaction. Fig. 5 shows the composition profiles in the particles. It is noted that the cross-interaction has a stronger effect on the y profile in $(Ti_xNb_{1-x})(C_yN_{1-y})$ than on the x one. As can also be seen there is a significant multicomponent diffusion effect on the interstitial profiles in the austenite and the particle, but a weaker effect

TABLE 4
Lattice Parameters of MX

MX	Lattice Parameters (Å)	Reference
NbC	4.470	Nagakura 22
TiC	4.326	Nagakura 22
NbN	4.392	Goldschmidt 23
TiN	4.244	Brauer 24

on the interface motion (particle size), since this is controlled by the slow-diffusing Ti and Nb.

Fig. 6 shows STEM and TEM micrographs of growth of precipitates in the present alloys during 1 to 20 s (5). At 1 s particles as small as 1 nm were observed which is not much greater than particles at the critical radius r_c. Fig. 7 gives a comparison of the predicted and observed average diameter which is salutary to the multicomponent diffusion model. Similar results were obtained for a range of alloys (5).

Discussion

The phase transformation modeller in complex multicomponent systems must always confront a decision whether or not to include diffusion cross-effects. One might conclude in the first instance that when the matrix concentrations are very low these effects can be neglected on account of relations (13) where the effect is seen to vanish with concentration. However, it must be noted that the interaction parameters are related to free energies of formation and therefore are not independent of the solubilities. Indeed they tend to lie in an inverse relation of magnitudes as can be seen from the present and the Fe-C-Mn example quoted, and in the accompanying papers by Young et al. (26) and Buchmayr and Kirkaldy (27). We are therefore inclined to conclude that for every multicomponent problem, where the solubility limit is exceeded, including those associated with alloy oxidation or sulfidation, the matrix cross-effects must be included for predictive precision. This, of course, is predicated upon the assumption that other parameters are sufficiently accurately known and that accurate modelling has a technological significance.

References

1. G.R. Purdy, D. Weichert and J.S. Kirkaldy, Trans. Met. Soc. AIME, 230, 1025 (1964).
2. L. Onsager, Ann. N.Y. Acad. Sci., 45, 241 (1945).
3. J.S. Kirkaldy and D.J. Young, Diffusion in the Condensed State, Institute of Metals, London (1987).
4. J.S. Kirkaldy, Met. Trans. 4, 2327 (1973).
5. H. Zou and J.S. Kirkaldy, submitted to Met. Trans. A (1990).
6. M. Hillert and L.-I. Staffansson, Acta Chem. Scand. 24, 3618 (1970).
7. J.S. Kirkaldy and G.R. Purdy, Can. J. Phys. 40, 208 (1962).
8. W.E. Voice and R.G. Faulkner, Met. Sci. 68, 411 (1984).

9. T.H. Johansen, N. Christensen and B. Augland, Trans. TMS-AIME, 239, 1651 (1967).
10. R.P. Smith, Trans. TMS-AIME, 236, 220 (1966).
11. T. Sharaiwa, N. Fujino and J. Murrayama: Trans. ISIJ, 10, 406 (1970).
12. K. Balasubramanian and J.S. Kirkaldy, McMaster University, unpublished research (1989).
13. R.P. Smith, TMS-AIME, 220, 476 (1964).
14. L.S. Darken, R.P. Smith and E.W. FIler, Trans. AIME, 191, 1174 (1951).
15. S.J. Moll and R.W. Ogilvie, Trans. TMS-AIME, 215, 613 (1959).
16. S. Kurokawa, J.E. Ruzzante, A.M. Hey and F. Dyment, Met. Sci. 17, 433 (1983).
17. C. Wagner, Thermodynamics of Alloys, Addison-Wesley, Reading Mass., (1952)
18. J.C. Greenband; 151, 208, 986 (1971), vol. 208, p. 986.
19. K. Balasubramanian and J.S. Kirkaldy: CALPHAD, 10, 187 (1986).
20. K. Balasubramanian and J.S. Kirkaldy: Can. Met. Quart. 28, (1989).
21. W. Roberts and A. Sandberg, Swedish Institute for Metals Research Report No. IM-1489, p. 301, Stockholm (1980).
22. S. Nagakura and S. Oketani, Trans. ISIJ, 8, 265 (1968).
23. H.J. Goldschmidt, Interstitial Alloys, Plenum Press, New York (1967).
24. G. Brauer and J. Jander, Z. Anorg. Chem. 270, 160 (1952).
25. M.C. Inman and H.R. Tipler, Met. Rev. 8, 134 (1963).
26. D.J. Young and J.S. Kirkaldy, this volume.
27. B. Buchmayr and J.S. Kirkaldy, this volume.

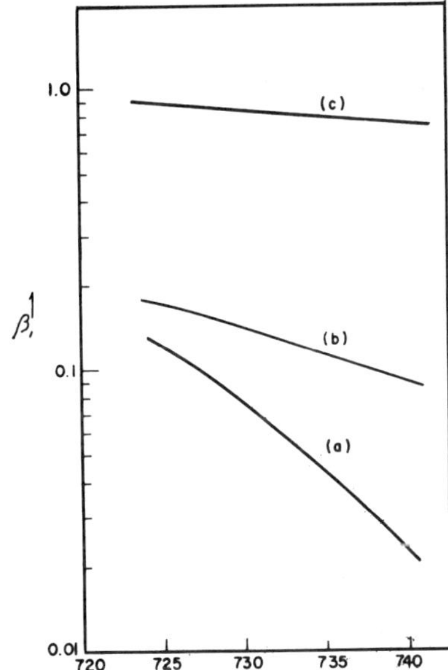

Fig. 1 Calculated variation of rate parameter β_1 with temperature: (a) complete calculation for the grain boundary reaction; (b) assuming a manganese constitutional effect, but neglecting ternary-diffusion effects ($D_{12} = 0$)p; (c) assuming that the alloy contains no manganese. After Purdy et al (1).

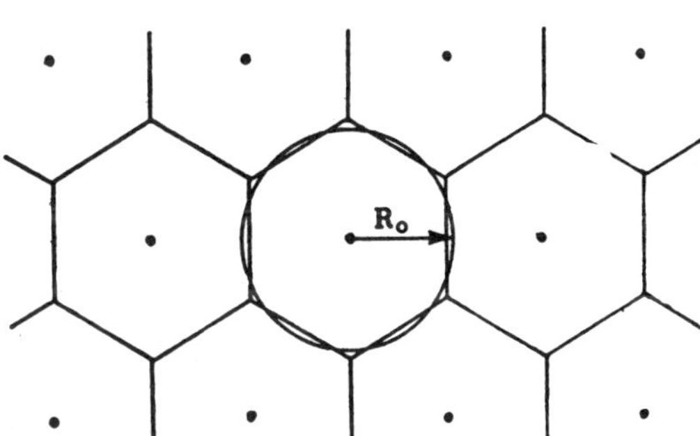

Fig. 2 The Wigner-Seitz array of equi-sized and spacing nuclei surrounded by cells from which they precipitate.

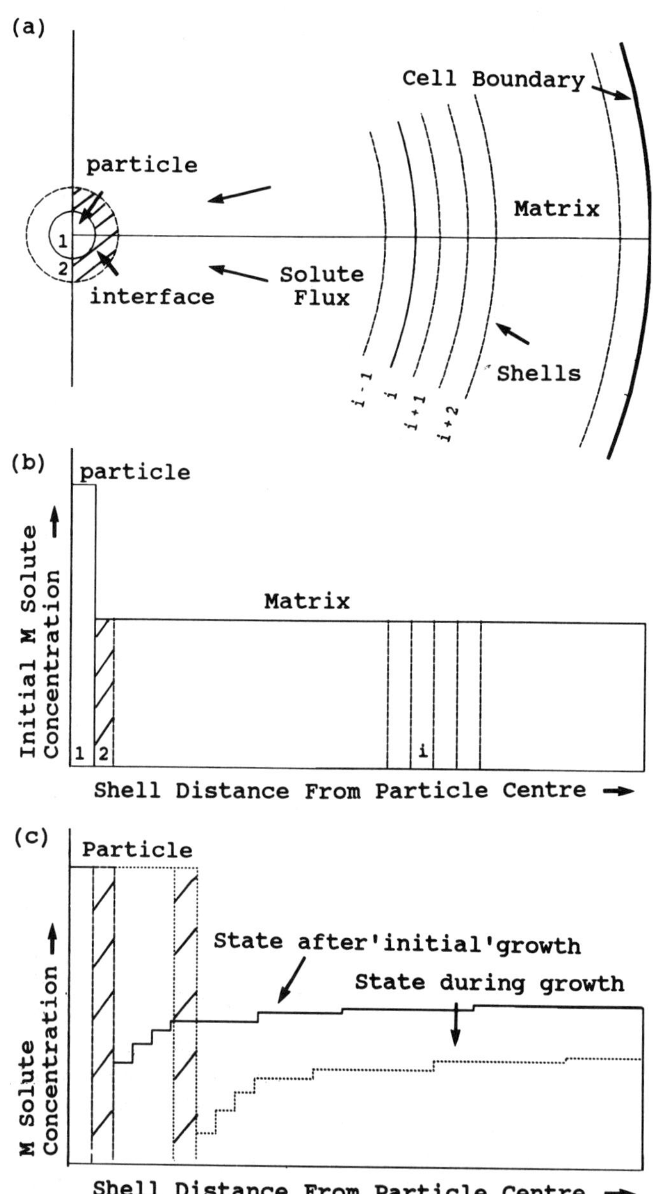

Fig. 3 Schematic of computer simulated growth of spherical carbonitride particle; (a) geometry of the shell; (b) initial solute concentration profile (c) solute concentration profile during growth.

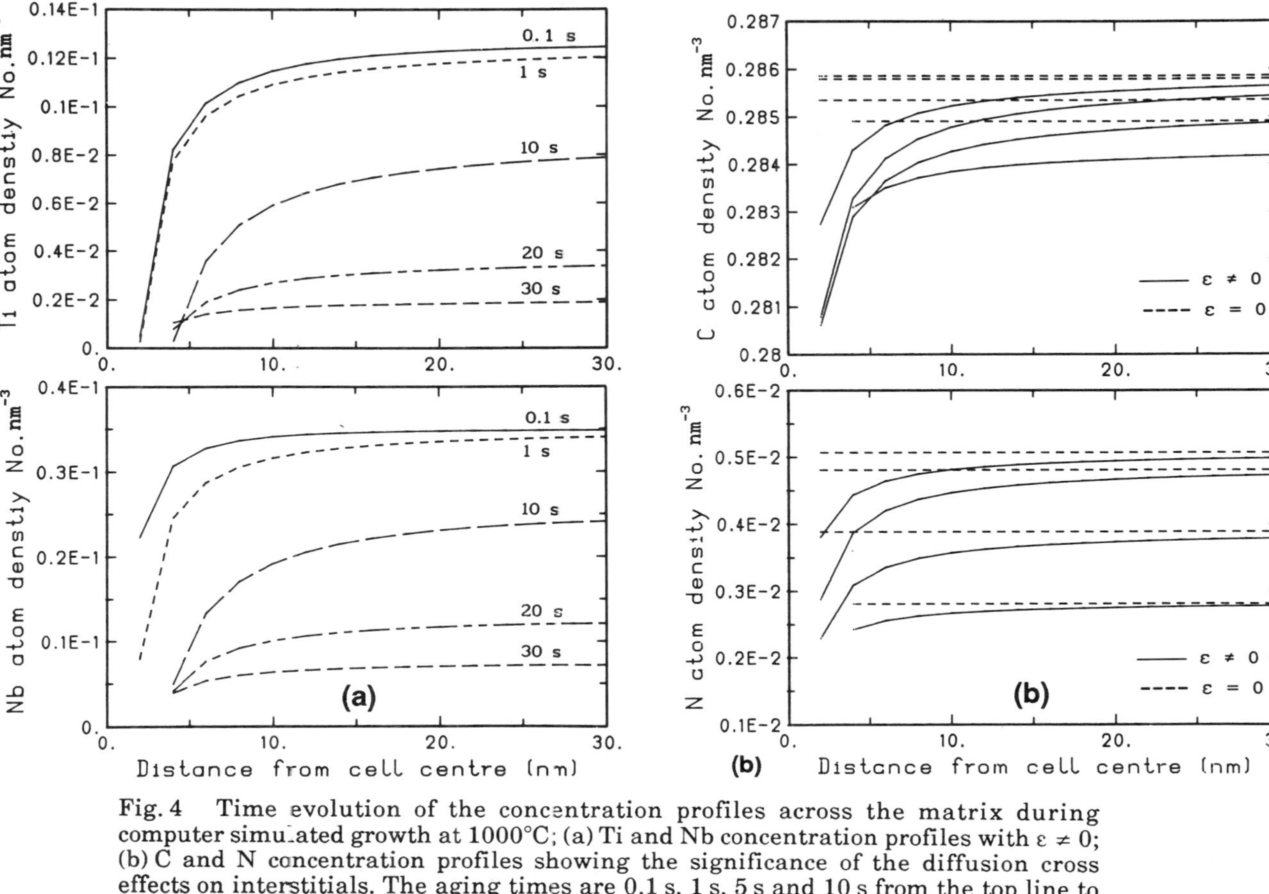

Fig. 4 Time evolution of the concentration profiles across the matrix during computer simulated growth at 1000°C; (a) Ti and Nb concentration profiles with $\varepsilon \neq 0$; (b) C and N concentration profiles showing the significance of the diffusion cross effects on interstitials. The aging times are 0.1 s, 1 s, 5 s and 10 s from the top line to the bottom.

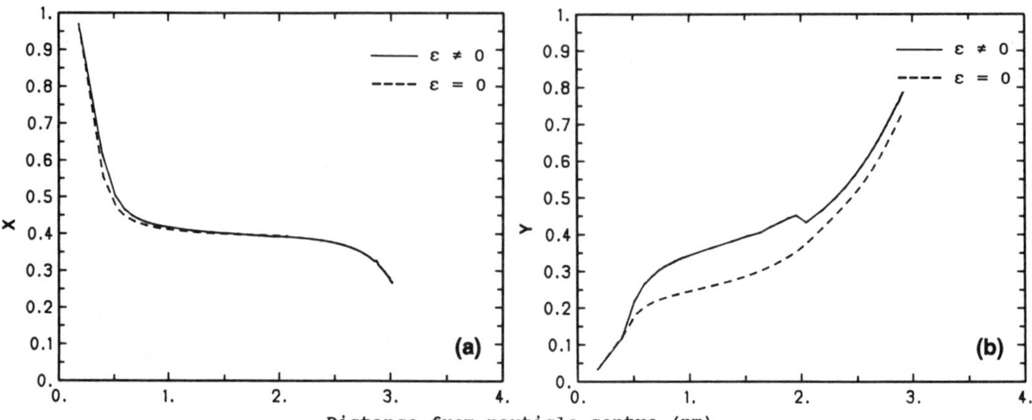

Fig. 5 Predicted composition profiles in the complex particles at 1000°C (i.e., x and y in $(Ti_xNb_{1-x})(C_yN_{1-y})$), illustrating the effects of cross-interaction on the inner composition profiles of the particles. (a) x profile; (b) y profile.

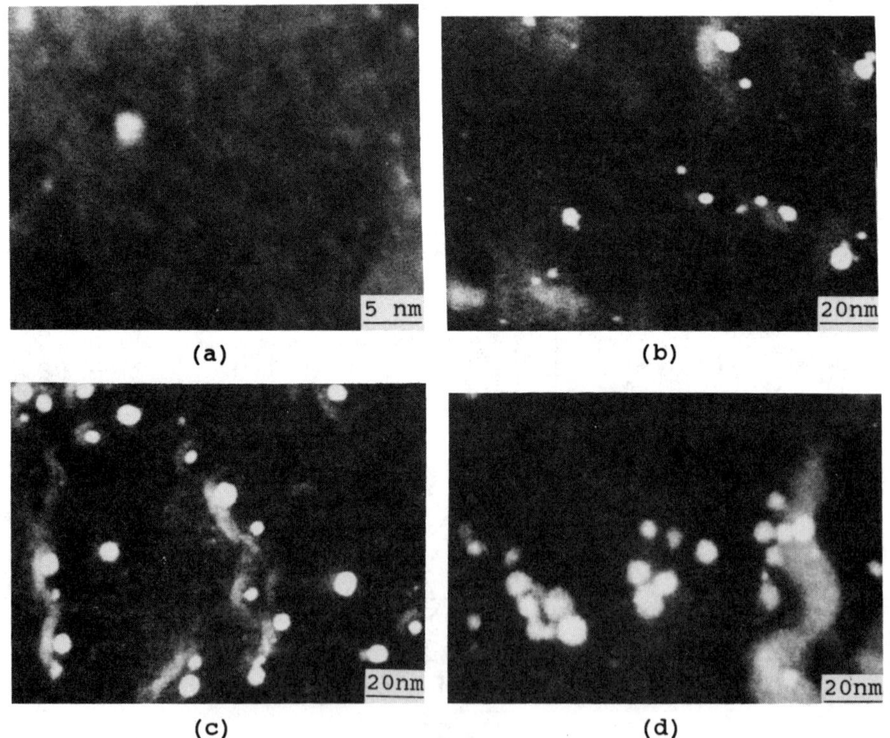

Fig. 6 STEM dark field micrographs of (TiNb)(CN) precipitates showing growth for various ageing times at 1000°C; (a) 1 s, (b) 5 s, (c) 10 s, (d) 15 s.

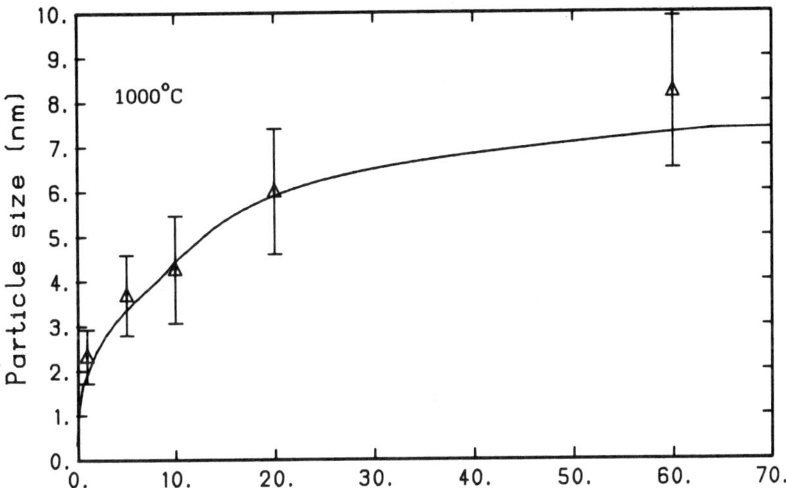

Fig. 7 Variation of average carbonitride particle sizes with time (pre-coarsening), solid line predicted from model, triangle - experimental.

A first-order solute-drag theory for effects of alloying elements on the growth of proeutectoid ferrite in steels

G.R. Purdy
Department of Materials Science and Engineering, McMaster University, Hamilton, Ontario, Canada, L8S 4M1

Y.J.M. Brechet
L.T.P.C.M./E.N.S.E.E.G., Domaine Universitaire, St-Martin d'Hères, Cedex, France

Abstract

A simple model is developed to describe the effects of alloying elements on the proeutectoid ferrite transformation in steels, for cases where the interfacial velocity is comparable to D_{2i}/δ, the quotient of the substitutional interfacial diffusion coefficient and the interfacial thickness. The model allows for the loss of local equilibrium for the substitutional alloying element with increasing interface velocity. Local equilibrium is maintained for the interstitial solute, carbon. The treatment allows for a transition from local equilibrium to "paraequilibrium" conditions, and for a "solute drag" effect under certain conditions. The shape stability of the transformation front is considered.

Introduction

In this contribution, we develop a simple model for the effects of alloying elements on the rate of growth of proeutectoid ferrite in steels. The model has its roots in the local equilibrium description of ternary phase transformation kinetics [1,2,3,4] and in the theory of solute drag forces on grain boundaries and interphase boundaries [5,6,7,8]. The need for such a development has been expressed repeatedly by Aaronson and his co- workers [9,10,11], and, indeed, elements of the current treatment are present, albeit in a qualitative sense, in their publications, as well as in an early and comprehensive treatment by Hillert [6]. It is the Hillert model which is explored and expanded in this work. We will, however, make some new assumptions concerning the rate-determining processes for the transformations.

We will see that the local equilibrium treatment of ferrite growth kinetics can be simply extended to include an interphase boundary "phase" characterised by a specific interaction with the two solutes. The boundary phase is modelled, for simplicity, as a square well. Departures from local equilibrium are then a consequence of limited diffusion within and ahead of the boundary phase, in precise analogy with the treatment of solute drag in binary solid-solid transformations. Subsequently, the loss of substitutional solute enrichment in the boundary phase with increasing velocity (a transition to "paraequilibrium") and the shape stability of the transformation interface in the solute drag regime are considered.

Application to Fe-C-X (where X is a substitutional austenite stabiliser)

As in previous treatments [1,3], we consider two solutes, one highly mobile (the interstitial solute, carbon, 1,) and the other a slower diffusing substitutional solute such as manganese, 2. Based on the local equilibrium hypothesis, the relevant isothermal section of the phase diagram can be divided into two regions: one in which the partition of the alloying element is required, and one in which a partitionless transformation is expected. These regions are separated by the locus of corners, c, of triangles such as a b c, Figure 1, where ab is a tie-line, bc an isoactivity line for carbon in the austenite phase, and ac a line of constant concentration of element 2. We will restrict discussion to transformations within the region of higher supersaturation, for which a partitionless transformation is often observed, and for which carbon diffusion is thought to be rate determining, provided only that the interface is sufficiently mobile. Solute segregation to the boundary is characterised by two equilibrium distribution coefficients

$$k_1^{i/\alpha} = C_{1i}/C_{1\alpha} \quad ; \quad k_2^{i/\alpha} = C_{2i}/C_{2\alpha} \tag{1}$$

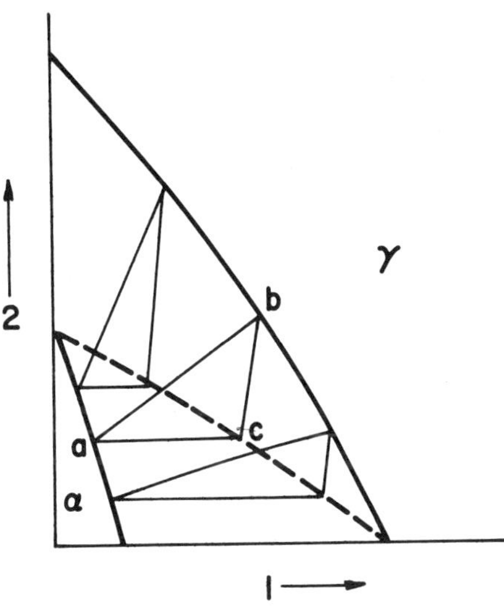

Figure 1 Isothermal section of the equilibrium diagram Fe-C-X for an austenite-stabilizing element X (2). A series of tie-lines (ab) are shown and the "envelope at zero partition" (broken).

Similarly, the equilibrium distributions between α and γ phases are given by

$$k_1^{\alpha/\gamma} = C_{1\alpha}/C_{1\gamma} \quad ; \quad k_2^{\alpha/\gamma} = C_{2\alpha}/C_{2\gamma} \tag{2}$$

thus

$k_1^{i/\gamma} = k_1^{i/\alpha} \cdot k_1^{\alpha/\gamma}$ and $k_2^{i/\gamma} = k_2^{i/\alpha} \cdot k_2^{\alpha/\gamma}$ are determined.

The interfacial diffusion coefficients D_{1i} and D_{2i} are thought to be much greater than those for the parent austenite, especially for the substitutional solute, 2. The schematic distribution of figure 2 is for a velocity low enough that full local equilibrium applies; there is no significant dissipation within the interface, and the effective interfacial carbon concentration $C^*_{1\gamma}$ is found at the corner, c, of the appropriate triangle containing the no- partition tie-line. This carbon content enters the kinetic equations as a rate determining quantity [1-4]. The basic requirement, which will be carried forward to non-equilibrium interfacial conditions, is that the activity of the mobile solute, 1, be slowly varying through the interface region.

At higher velocities, the "spike" in the austenite is lost (D_{22}/v less than about one nm), as a result of the relative slowness of substitutional diffusion in austenite. The effect of this loss on the effective interfacial concentration $C^*_{1\gamma}$ is negligible, however, as this value is still determined by the intersection of the relevant isoactivity line with $C_2 = C_{20}$, the bulk concentration of substitutional alloying element in the austenite.

At still higher velocities, diffusion within the interface is expected to become important, and the concentration difference ΔC^i_2 will develop across the boundary, as suggested by figure 3.

$$\Delta C_2^i = k_2^{i/\alpha} C_{20} \{1 - e^{-v\delta/D_{2i}}\} \tag{3}$$

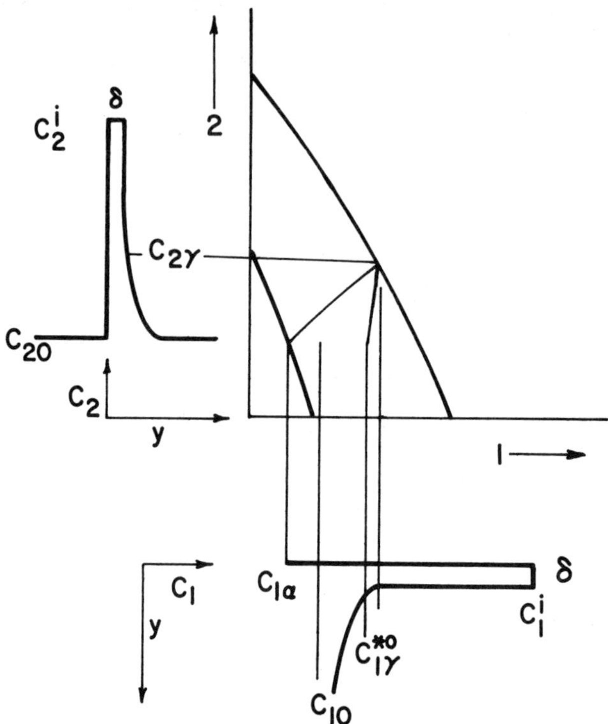

Figure 2 The local equilibrium boundary condition, for an alloying element, (2) which co-segregates with carbon (1), to the transformation interface.

A corresponding interfacial variation in the concentration of component 1 will develop, according to

$$\Delta C_1^i = \varepsilon_{12}^i \Delta C_2^i \tag{4}$$

where ε_{12} describes the variation of the activity of component 1 with concentration of component 2 within the boundary. Note that the effective interfacial carbon concentration is changed somewhat from its local equilibrium value, $C^{*0}{}_{1\gamma}$, by an amount $\partial C^*{}_{1\gamma}$, and that the rate of transformation will therefore be reduced by this variation of concentration within the interface. This is the first indication of a solute drag effect, due to the variation of the composition of substutional solute across the interface, and the (implicit) condition of a consequent variation of composition of the carbon concentration through the interface. In effect, at this stage, we have allowed the loss of local equilibrium at the front of the transformation interface, but retained the condition at the back of the interface, the α/i plane.

The idea that equilibrium is first lost (with increasing velocity) in the parent phase, then at the front of the transformation interface, finds support in recent studies of discontinuous precipitation in binary alloys [12], which suggest that the product phase remains closely in equilibrium with the high-conductivity transformation front.

In order to pursue the solute drag effect in this system, we must next account for a loss of equilibrium at the α/i plane. This may be done, following Hillert and Sundman [7], by assigning finite slopes to the sides of the solute-boundary interaction profile, or, more simply, by taking as interfacial concentrations substitutional solute contents at $y = \delta/4$ and $y = 3\delta/4$. Indeed, this approximation yields results rather close to those of the full computation. Then the alloying element concentration difference across the boundary becomes

$$\Delta C_2^i = k_2^{i/\alpha} C_{20} \{ e^{-3v\delta/4D_{2i}} - e^{-v\delta/4D_{2i}} \} \tag{5}$$

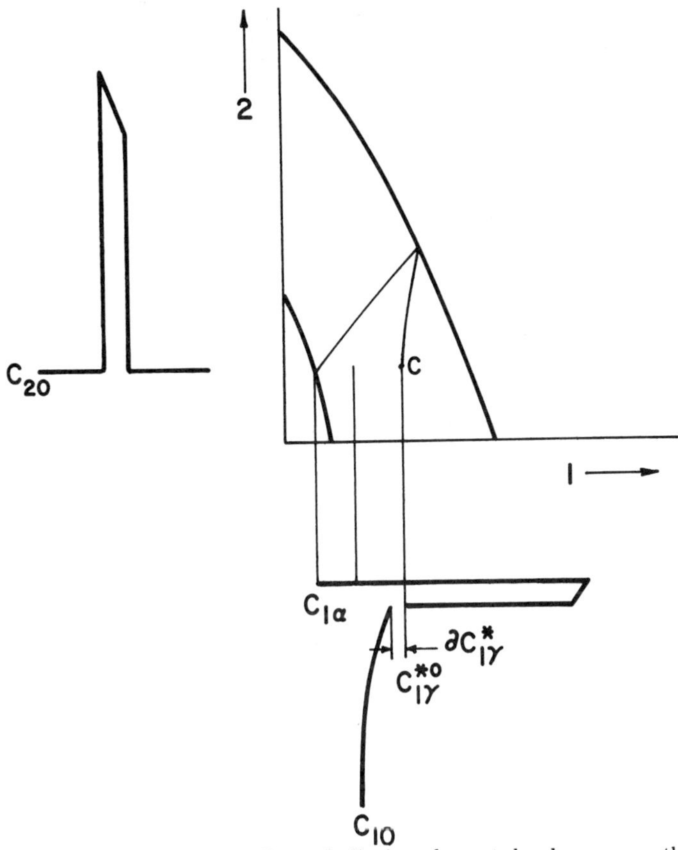

Figure 3 When a concentration gradient of alloying element develops across the interface, a corresponding carbon gradient results, and the local equilibrium carbon content in austenite is reduced.

As shown by Figure 4, this function allows for the disappearance of the alloying element spike within the boundary with increasing velocity. When ΔC^{i}_{2} becomes sufficiently small (when $v\delta/D_{2i}$ is greater than about 10) a transition to the "para" transformation mode is complete. Intermediate velocity states correspond to increasing carbon activities in the γ phase, and consequent higher values of the interfacial carbon content. We can reproduce this behaviour with a new term

$$\partial C^{+}_{1\gamma} = (C^{p}_{1\gamma} - C^{*0}_{1\gamma})\,\{1 - e^{-3v\delta/4D_{2i}}\} \tag{6}$$

where $\partial C^{+}_{1\gamma}$ is the increment in the interfacial carbon content due to this further departure from local equilibrium. The maximum value of $\partial C^{+}_{1\gamma}$, that is, the difference between the paraequilibrium interfacial carbon concentration in austenite, $C^{p}_{1\gamma}$, and its local equilibrium counterpart, $C^{*0}_{1\gamma}$, is indicated in Figure 5.

Thus, velocity-dependent effects of opposite sign are predicted. The first is a drag term, due to the interaction of the alloying element diffusion field within the boundary with that of carbon; the second is an accelerating term, related to the gradual loss of equilibrium at the α/i plane, and culminating in the loss of alloy element diffusion and a true paraequilibrium boundary condition. For the system, Fe-C-Mn, the Mn-C interaction is modest, and the loss of α/i equilibrium is already significant when ΔC^{i}_{2} is a maximum, (e.g. at $v\delta/D_{2i} \sim 3$). We therefore expect no net diminution of rate from the full local equilibrium value, and a steady acceleration of the transformation rate with increasing departure from equilibrium within the interface.

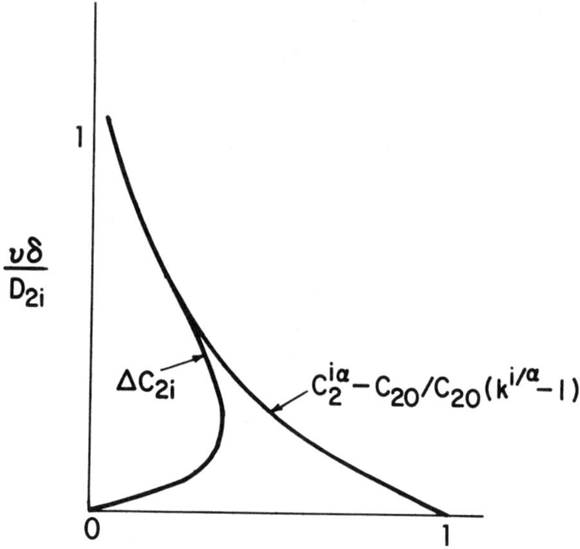

Figure 4 The velocity dependence of the substitutional element interfacial concentration difference δC_{2i}, and the loss of local equilibrium at the i/α plane.

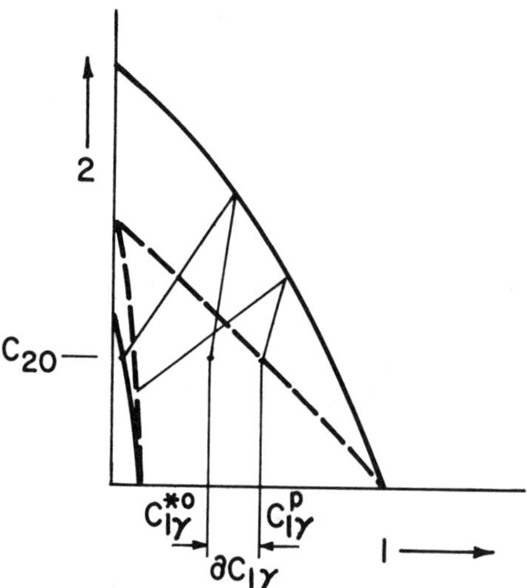

Figure 5 The maximum value of $\delta C_{1\gamma}{}^{+}$ occurs when the full paraequilibrium condition is attained. Paraequilibrium boundaries are shown as broken lines.

<u>Application to Fe-C-X, where X is a substitutional ferrite stabilising carbide forming alloying element</u>

The Fe-C-Mo system, Figure 6, is characterised by nearly horizontal tie-lines in the region of present interest. Thus, the local equilibrium spike of component 2 in the parent phase will not be significant; indeed, for most transformation rates, and alloy systems, such spikes will not exist [4,9]. Using a construction analogous to that of Figure 3, we find that the effective interfacial carbon concentration is again reduced by a drag term, proportional to the alloying element concentration difference across the boundary

$$\partial C^*_{1\gamma} = \frac{\varepsilon^i_{12} C_{20} k_2^{i/\alpha}}{k_1^{i/\alpha} \cdot k_1^{\alpha/\gamma}} \{e^{-3v\delta/4D_{2i}} - e^{-v\delta/4D_{2i}}\} \tag{7}$$

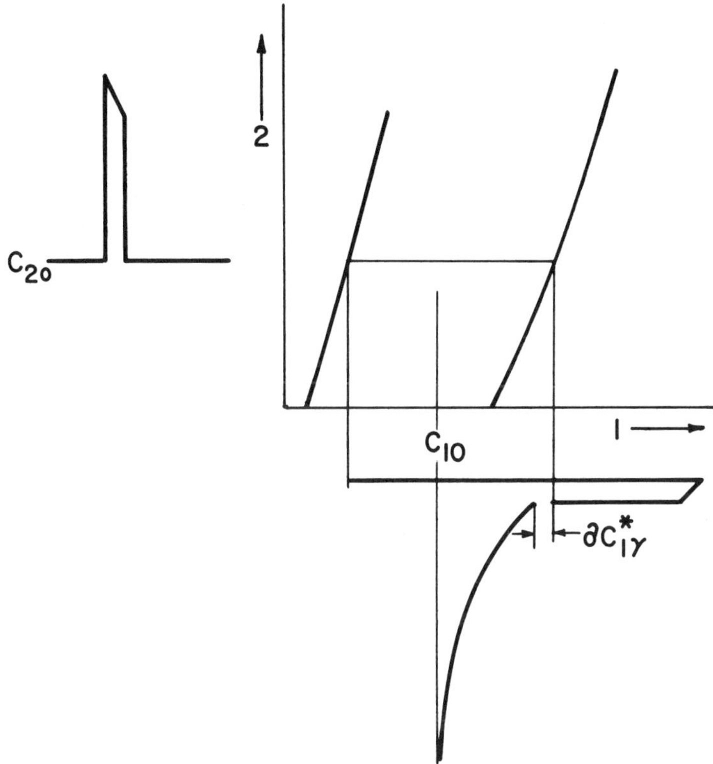

Figure 6 The solute drag effect for a ferrite stabilizing alloying element, as in figure 3.

This is shown in Figure 6. This drag term is not offset by a simultaneous accelerating term due to the loss of equilibrium at the α/i plane, and it is therefore expected to be manifested as a diminution of growth rate at temperatures where the interfacial diffusion becomes significant. At higher values of v/D_{2i}, the solute drag term will be lost as the interface "outruns" the solute field, and a high velocity state is again established.

The Morphological Stability of a Transformation Interface Influenced by Substitutional Solute Diffusion

Consider next the possibility that substitutional solute flow parallel to the interface may act to destabilise a planar growth front. We restrict attention to the regime of lower velocities where the α/i plane is in approximate local equilibrium, but where the concentration profile of the substitutional solute in the parent phase is lost. (This is thought by some writers to be a rather common condition.) In treating the stability of the transformation interface, we begin by distinguishing two cases; one in which the velocity is decreased when the interfacial alloying element concentration is locally increased, ($F<0$, Case I), and one in which the velocity is positively correlated with local alloying element concentration, ($F>0$, or Case II); the latter situation can lead, under certain conditions to an inherently unstable interface.

The local boundary concentration, C^i_2 will be permitted to differ from the average boundary concentration, $\bar{C}^i_2 = k^{i/\alpha}{}_2 C_{20}$ (in the low velocity approximation). With Roy and Bauer [8], we write the flux along the boundary as a sum of convective and diffusive terms:

$$J^i_2 = -N C^i_2(x,t)\, v(x,t) \sin\theta + J_{20} \tag{8}$$

Then the equations of motion for the boundary are

$$\frac{\partial \xi}{\partial t} = \gamma \frac{\partial^2 \xi}{\partial x^2} + F(C^i_2 - \bar{C}^i_2) \tag{9}$$

$$\frac{\partial C^i_2}{\partial t} = N C^i_2 v \frac{\partial^2 \xi}{\partial x^2} - \frac{C^i_2 - \bar{C}^i_2}{\tau} + D^i_2 \frac{\partial^2 C^i_2}{\partial x^2} \tag{10}$$

where γ is the product of the specific interfacial free energy and an effective mobility, ξ is the fluctuation amplitude, and other variables of the problem are defined in Figure 7. The time τ characterises the rate of relaxation of local concentration fluctuations within the boundary.

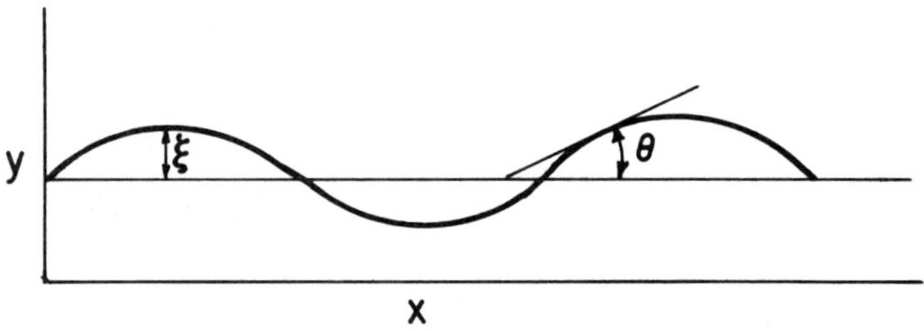

Figure 7 The co-ordinate system for the perturbation analysis of interfacial stability.

The set (9, 10) has eigenvalues $\omega(k)$ which are solutions of

$$\omega^2 + \omega\{\gamma k^2 + 1/\tau + D_{2i}\, k^2\} + \{\frac{\gamma k^2}{\tau} - F N \bar{C}^i_2 v_0 + \gamma D^i_2 k^4\} = 0 \tag{11}$$

where v_0 is the quasisteady velocity of the plane front. Then $\omega(k)$ gives the rate of exponential amplification or decay of a sinusoidal fluctuation of wavevector k.

We then find that for $F<0$, (Case I), the interface is stable against perturbations occasioned by lateral substitutional solute flow. However, for $F>0$, or Case II, the interface is unstable for a range of wavenumbers between k' and k'', the roots of

$$\gamma D^i_2 k^4 + \frac{\gamma k^2}{\tau} - F N \bar{C}^i_2 v_0 = 0 \tag{12}$$

The situation is illustrated in Figure 8. A maximum amplification rate is expected for the finite wavevector k_m.

This result can be interpreted in terms of the equilibrium diagrams of Figure 9, a,b, where the effect of an increase in the boundary composition C^i_2 is reflected as an increase in the local concentration in ferrite, $C_{2\alpha}$. For case I, Figure 9 a, the result will be to increase the 2 activity and decrease the carbon concentration in the interface region by $\partial C_{1\gamma}$. The local velocity will then be reduced. In contrast, the ferrite stabilising solute of Figure 9 b will locally increase the carbon activity and the interface velocity, corresponding to Case II, resulting in the possibility of an unstable growth pattern.

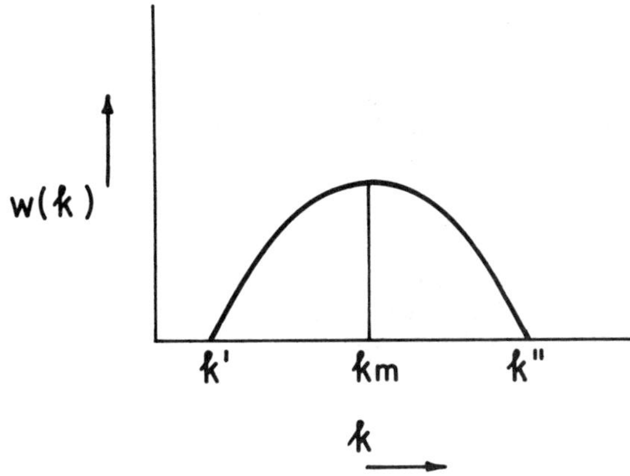

Figure 8 ω(*k*) as a function of *k*.

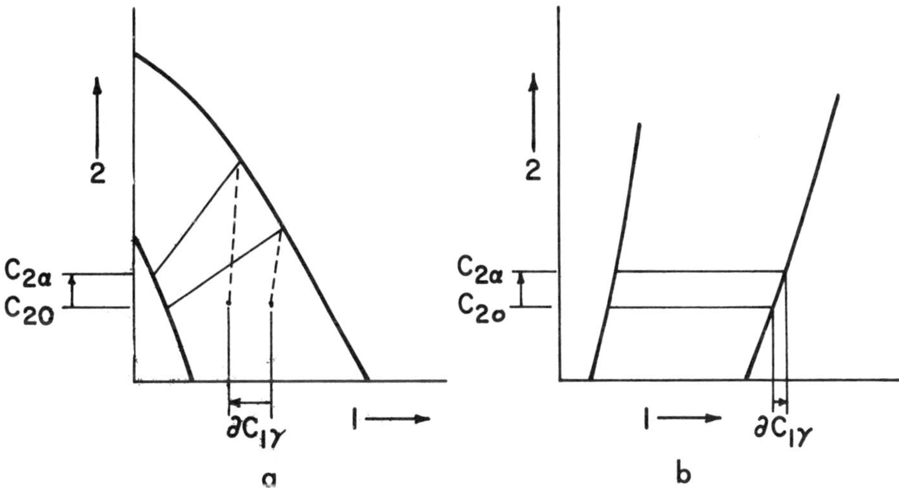

Figure 9 Lateral substitutional solute diffusion results in (a) a reduction in interfacial austenite carbon content with increased alloying element concentration, or (b) an increase in velocity with increased alloying element concentration.

Discussion

We find that the first-order treatment of solute diffusion within the transformation interface predicts or explains a number of phenomena which have heretofore eluded quantitative description. The key assumption in applying solute drag theory to the proeutectoid ferrite transformation interface is that the transformation rate is controlled by the diffusion of the interstitial solute, carbon, and that the interface mobility does not directly enter the rate of transformation. Then the dissipation of free energy due to solute diffusion within the interface, although it can be expected to produce a drag term which influences the interfacial mobility, will also not influence directly the transformation rate. In all cases, it is the carbon activity in the austenite immediately in front of the boundary which is the main determinant of growth kinetics, and we have seen that this quantity is sensitive to diffusive processes within the interface. For alloying elements such as manganese, which have rather weak negative interactions with carbon, and which stabilise austenite, the treatment suggests that there will be no real drag, or diminution of the growth rate due to manganese diffusion within the boundary; rather the weak drag effect will be overcome by an accelerating term due to the gradual transition to a paraequilibrium mode and an associated increase in the effective interfacial carbon concentration $C^*_{1\gamma}$.

In contrast, ferrite stabilisers which are also strong carbide formers will exert a real drag effect, as the cross-boundary alloying element gradient is matched by a carbon gradient, and the effective interfacial carbon concentration is reduced. The effect of this drag term will be to produce a "bay" in the isothermal transformation kinetic curves, as is commonly observed in steels containing, say, molybdenum and chromium. The effect will persist until the interface velocity is increased or the interfacial diffusion coefficient is reduce to allow the interface to outrun the solute, and a paraequilibrium state is established.

In this discussion, it has been convenient to speak of the solute drag effect as developing with increasing interface velocity, i.e. with increasing departures from the local equilibrium state. In fact, in the course of a typical isothermal transformation, the initial velocity will be higher, and the interfacial conditions closer to those predicted by the paraequilibrium model. The drag effects will likely influence growth kinetics at rather later stages of the transformation.

In keeping with the first-order nature of the treatment of solute effects, we have presented a first-order calculation of the stability parameters for the planar interface. as influenced by substitutional solute diffusion parallel to the boundary. We find that instability is expected for the case where the local solute concentration is positively correlated with the local velocity, as, for example, in the case of Fe-C-Mo alloys. With this in mind, we refer the reader to an unusual product of austenite decomposition called "wrinkled ferrite" [13-16]. This form of densely substructured ferrite is found in steels containing about 1 at% Mo, transformed at or near the temperatures of the bay (e.g. 600°C), and for interfacial orientation relationships which, by inference, lead to interface structures less coherent than the broad faces of Widmanstatten laths. We consider that it is possible that this microstructure, illustrated in the optical micrograph of Figure 10, results when the criteria for morphological instability are met, and that it results from an instability accompanying the lateral diffusion of molybdenum in the transformation interface.

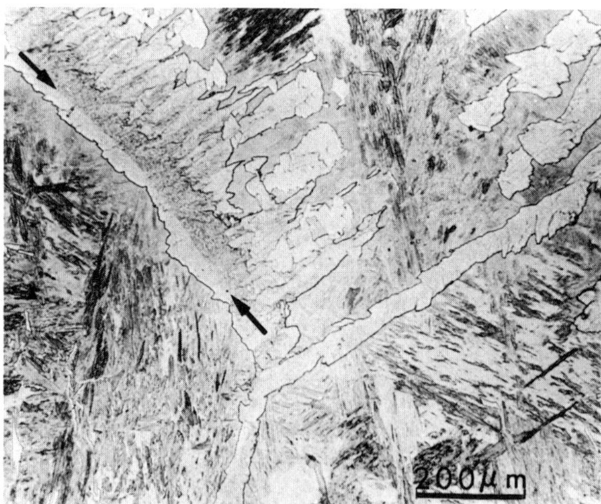

Figure 10 An optical micrograph, showing the formation of "wrinkled ferrite" on one side of a prior austenite grain boundary Fe-0.35% C-1% Mo, isothermally transformed at 650°C.

In conclusion, the simple theory given here is shown to be an extension of the local equilibrium description of ferrite growth in ternary steels. It predicts a transition, through a sequence of intermediate states, to the paraequilibrium mode of transformation. All of the transformation states are assumed controlled by the diffusion of the mobile solute, carbon; it is the interfacial carbon concentration in the austenite immediately adjacent to the transformation interface which is influenced by alloying element diffusion within the interface, and which is a principal determinant of the rate of ferrite growth on this model.

We have neglected many important questions related to interfacial structure and intrinsic mobility, (as do all treatments of solute drag phenomena,) but consider that the model is of interest in the sense that it yields results which are consistent with the general behaviour of transformation kinetics of alloy steels. In particular, the existence of a bay in the isothermal transformation kinetics and the occurrence of wrinkled ferrite structures in molybdenum steels can be understood.

Acknowledgements

This research was supported by the Natural Sciences and Engineering Research Council of Canada and the Ontario Centre for Materials Research.

References

1. J.S. Kirkaldy, Can. J. Phys., 36, 907 (1958).
2. M. Hillert, Internal report, Swedish Institute of Metal Research, (1953).
3. G.R. Purdy, D.H. Weichert and J.S. Kirkaldy, Trans. AIME 230, 1025 (1964)
4. D.E. Coates, Metall. Trans. 3, 1203, (1972).
5. J.W. Cahn, Acta Metall. 10, 789 (1962).
6. M. Hillert, in "The Mechanism of Phase Transformations in Crystalline Solids", p. 231, The Institute of Metals, London (1969).
7. M. Hillert and B. Sundman, Acta Metall., 24, 731 (1976).
8. A. Roy and C.L. Bauer, Acta Metall., 23, 957 (1975).
9. W.T. Reynolds, M. Enomoto and H.I. Aaronson, in "Phase Transformations in Ferrous Alloys", TMS-AIME Wanendale PA, p. 155 (1984).
10. K.R. Kinsman and H.I. Aaronson, in "Transformation and Hardenability in Alloy Steels", Climax Molybdenum Co., Ann Arbor, MI, p. 39 (1967).
11. H.I. Aaronson, M. Enomoto and W.T. Reynolds, in "Advances in Phase Transitions" (Eds. Embury and Purdy), Pergamon Press plc, p. 20 (1988).
12. K. Tashiro and G.R. Purdy, Metall. Trans. A, 20A, 1593 (1989).
13. A. Hultgren, Kungl. Svenska Veleuskapskademieus Handlingar 4 (3) (1953).
14. W. Pitch and A. Schraeder, unpublished research, see De Ferri Metallographia II, by A. Schraeda and A. Rose, pp. 367-368, Verlag Stahleisen m.b.H. Düsseldorf (1966).
15. G.R. Purdy, Acta Metall. 26, 487 (1978).
16. R.B. Brown, H. Badekas and G.R. Purdy, Metallography 16, 375 (1983).

Authors' Index

Subject Index